D1080858

Set-off
in the Construction Industry

Set-off

in the Construction Industry

Second Edition

Neil F. Jones Solicitors

**Blackwell
Science**

© 1991 by Neil F. Jones & Co., 1999 by Neil F. Jones Solicitors

Blackwell Science Ltd
Editorial Offices:
Osney Mead, Oxford OX2 0EL
25 John Street, London WC1N 2BL
23 Ainslie Place, Edinburgh EH3 6AJ
350 Main Street, Malden
  MA 02148 5018, USA
54 University Street, Carlton
  Victoria 3053, Australia
10, rue Casimir Delavigne
  75006 Paris, France

Other Editorial Offices:

Blackwell Wissenschafts-Verlag GmbH
Kurfürstendamm 57
10707 Berlin, Germany

Blackwell Science KK
MG Kodenmacho Building
7–10 Kodenmacho Nihombashi
Chuo-ku, Tokyo 104, Japan

The right of the Author to be identified as the
Author of this Work has been asserted in
accordance with the Copyright, Designs and
Patents Act 1988.

First edition published by The Chartered Institute
of Building 1991
Second edition published by Blackwell Science
1999

Set in 10.5/12.5 pt Palatino
by DP Photosetting, Aylesbury, Bucks
Printed and bound in Great Britain by
MPG Books Ltd, Bodmin, Cornwall

The Blackwell Science logo is a trade mark of
Blackwell Science Ltd, registered at the United
Kingdom Trade Marks Registry

DISTRIBUTORS

Marston Book Services Ltd
PO Box 269
Abingdon
Oxon OX14 4YN
(*Orders:* Tel: 01235 465500
         Fax: 01235 465555)

USA
Blackwell Science, Inc.
Commerce Place
350 Main Street
Malden, MA 02148 5018
(*Orders:* Tel: 800 759 6102
            781 388 8250
         Fax: 781 388 8255)

Canada
  Login Brothers Book Company
  324 Saulteaux Crescent
  Winnipeg, Manitoba R3J 3T2
  (*Orders:* Tel: 204 837-2987
           Fax: 204 837-3116)

Australia
  Blackwell Science Pty Ltd
  54 University Street
  Carlton, Victoria 3053
  (*Orders:* Tel: 03 9347 0300
           Fax: 03 9347 5001)

A catalogue record for this title is available
from the British Library

ISBN 0-632-04824-7

Library of Congress
Cataloging-in-Publication Data
is available

For further information on
Blackwell Science, visit our website:
www.blackwell-science.com

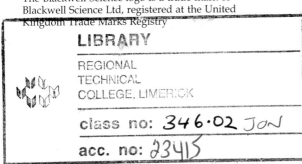

CONTENTS

Contents

PREFACE

In the famous set-off case of *Gilbert-Ash (Northern) Ltd* v. *Modern Engineering (Bristol) Ltd*, the late Lord Denning, Master of the Rolls sitting in the Court of Appeal, uttered the truism:

'There must be a "cash flow" in the building trade. It is very lifeblood of the enterprise.'

Nowhere is the significance of cash flow demonstrated more strikingly than in the realm of set-off. Who is to hold the money pending the resolution of disputes between contracting parties in arbitration or litigation? In recent years this question has occupied much of the time of the courts and draughtsmen of standard forms of subcontract in use throughout the construction industry.

The first edition of this book was published in 1991. Since then the subject of set-off has become increasingly important. Sir Michael Latham in his final report, *Constructing the Team*, published in 1994, proposed changes to the law of set-off and the introduction of a statutory adjudication procedure. The Housing Grants, Construction and Regeneration Act 1996, Part II of which came into force on 1 May 1998, enacted a framework of measures applying to construction contracts which also has an effect on set-off. The Joint Contracts Tribunal (JCT) and others promptly published amendments to their standard form contracts to conform with the Act. Added to this, the courts have had occasion, since publication of the first edition, to rule on various issues relevant to set-off as it affects the construction industry. Finally, the rules of court have been reformed as a result of the Woolf report, changing the grounds and procedure for summary judgment and interim payment applications. All of these matters are covered in this work.

The first edition was the result of a collaboration between Neil Jones, Jeffrey Brown, Kevin Barrett, Phillip Harris and Simon Baylis, all lawyers at Neil F. Jones & Co. The preparation of the second edition was largely undertaken by Kevin Barrett.

We have received assistance from many sources, not the least being clients who consult us with real problems with which to grapple. We are indebted to them. The task of re-typing the script (from scratch due to a missing disk) fell initially to Katrina Bevan. Subsequently, the major amendments were undertaken by Jacki Lovell. We are indebted to them for their endeavour and their patience.

Kevin Barrett
Partner
Neil F Jones Solicitors
3 Broadway
Broad Street
Birmingham
B15 1BQ
March 1999

GLOSSARY OF TERMS

Blue Form – The non-nominated (domestic) subcontract for use with JCT 63 (not to be confused with the FCEC Blue Form)

construction contract(s) – construction contract(s) as defined in the HGCRA

CPR – The Civil Procedure Rules which came into effect on 26 April 1999

DOM/1 – Standard Form of Domestic Subcontract

FASS – Federation of Specialist Subcontractors

FCEC Blue Form – Federation of Civil Engineering Contractors Blue Form of Subcontract

Green Form – The FASS nominated subcontract for use with JCT 63

HGCRA – The Housing Grants Construction and Regeneration Act 1996, Part II

IN/SC – BEC Domestic Subcontract IN/SC 1985 Edition

ICE – Institution of Civil Engineers

JCT – the Joint Contracts Tribunal

NAM/SC – JCT Standard Form of Named Subcontract Conditions

NSC/C – JCT Standard Form of Nominated Subcontract Conditions 1990

NSC/4 – JCT Standard Form of Nominated Subcontract NSC/4

the Scheme – the Scheme for Construction Contracts as defined in the HGCRA

CHAPTER ONE
THE NATURE AND HISTORY OF SET-OFF

1.1   Introduction

In 1954 Mrs Bozena Hanak retained a builder to carry out certain work to her house. Mrs Hanak was dissatisfied with the work and sued the builder for damages for breach of contract. Her claim was put in the sum of £266. The builder counterclaimed in respect of extra work and damages caused by Mrs Hanak's refusal to allow a workman into the house and her trespass against his tools.

These simple but commonplace facts ultimately gave rise to the leading modern analysis of the historical context of set-off – *Hanak* v. *Green* (1958) – because the builder asserted that he could set off his counterclaims against the claims made by Mrs Hanak. The analysis by the Court of Appeal, which agreed with the builder's assertion, resolved the dispute between Mrs Hanak and her builder. However, the problem of set-off continues to exercise the construction industry where there are claims and counterclaims between, say, an employer and a main contractor or between a main contractor and a subcontractor, because the question continues to arise as to whether one can be set off against the other.

The usual form of the main contract dispute is that the contractor seeks payment of the value of work done or reimbursement of direct loss and expense incurred, but the employer counters that the contractor failed to complete on time, or committed some other breach (such as the execution of non-conforming work), and that the resulting loss to the employer extinguishes all or part of the main contractor's claim. Additionally, counterclaims may arise from events unrelated to the contract under which payment is alleged to be due. Payment may be due from the employer to the contractor under one contract but the employer may wish to avoid payment by using the money due to compensate himself for unmet claims under another separate contract with the contractor, or as a result of a negligent or other tortious act on the part of the contractor which has caused the employer loss.

Disputes between main contractors and their subcontractors take a similar form to those arising out of the main contract relationship. In these circumstances, may that paying party set off any counter-claim or are there limits? In fact, as explained later, there are limits arising from the law of set-off, in some cases the express contractual terms and latterly the effects of the Housing Grants Construction and Regeneration Act 1996 (HGCRA).

The starting point for any consideration of issues concerning set-off is the law of set-off itself which consists of a series of principles that, where applicable, permit one person to withhold payment of money due to another.

The current state of the law of set-off is very much the product of history. It is complex and requires careful analysis in each case. In *Axel Johnson Petroleum AB* v. *MG Mineral Group AG* (1992) Leggatt LJ criticised it as unsatisfactory:

> 'The state of the law is unsatisfactory that allows a set-off at law of debts which are liquidated, even if unconnected, and in equity of debts which are connected, even if unliquidated but not a set-off of debts which are both unliquidated and unconnected.'

The law of set-off as it affects the construction industry operates subject to the same unsatisfactory state of the law. It is also subject to the effects, where they apply, of a variety of express contractual provisions and latterly the provisions of the HGCRA.

Under a contract for building and engineering work there may be cause for complaint about the performance of the work during the course of execution or after completion of the work. The most obvious and logical remedy available to the complainant is to deduct from payments otherwise due to the builder the sum of money which corresponds to the value attributed to the subject matter of the complaint.

This act of deduction is known variously as set-off, cross-claim, counterclaim or contra charge. However, some of these terms can carry different meanings. So far as legal terminology is concerned, the relevant expressions are counterclaim and set-off, but it is important to appreciate that set-off has a narrower meaning than counterclaim. All set-offs are counterclaims but not all counter-claims are set-offs. This point is dealt with in more detail later.

It must be said at once, however, that the importance of the distinction between a counterclaim which is also a set-off and a counterclaim which is not a set-off, cannot be over emphasised. Its importance lies in cash flow. If a counterclaim amounts also in law to a set-off, the party raising the set-off can withhold the value of the set-off from money otherwise due to the other party. If the counterclaim does not amount also in law to a set-off, the party raising the counterclaim has no general right in law to withhold the amount of the counterclaim from money otherwise due to the other party. The result is that the other party's claim must be paid in full, with the counterclaim being dealt with by way of a separate claim. The potential hardship which this could cause was to some extent ameliorated by the former Rules of the Supreme Court which in certain circumstances allowed the execution of a judgment on a claim to be stayed pending the resolution of the counterclaim, e.g. Order 14 rule 3 (2) and Order 47 rule 1 (see Chapter 6 for the position under the new Civil Procedure Rules).

Cash flow is the key in most of the cases, generally involving plaintiffs seeking to use summary procedures to release the cash. Prior to the HGCRA there were two main summary devices provided for in the Rules of the Supreme Court which could be employed to obtain money quickly:

(1)   under Order 14 and Order 14A for summary judgment without a trial of the action (superseded by Part 24 CPR)
(2)   under Order 29 for an interim payment on account (superseded by Part 25 CPR).

The HGCRA has introduced a new dispute resolution process called adjudication, which is likely to become the principal summary device for resolving set-off disputes and may therefore diminish the importance of the High Court summary procedures at least as the first stage of seeking relief. The two procedures provided by the High Court, and their limitations, are outlined in Chapter 6. Adjudication is dealt with in Chapter 3.

Despite the HGCRA the general common law position remains relevant to all contractual relationships; for example, between employer and main contractor, between main contractor and subcontractor and supplier, between subcontractor and sub-subcontractor and supplier, and so on. In this book therefore, the general common law position is explained in some detail, as is the background to the development of express set-off provisions limiting the contractor's rights of set-off in subcontracts issued by the Joint Contracts Tribunal (JCT), and indeed the background to and the effect of the HGCRA.

It is important to bear in mind that prior to the HGCRA the JCT forms of main contract did not expressly limit the employer's right to set-off. For example, in the case of *C. M. Pillings & Co Ltd* v. *Kent Investments Ltd* (1985) the plaintiffs were contractors to the defendants for extension work on a house and the provision of an indoor swimming pool. The contract was the JCT Fixed Fee Form of Prime Cost Contract 1967 Issue (Revised October 1976) which, like JCT 63 and JCT 80, contains no express terms dealing with set-off.

The architect issued a certificate, the accuracy of which was challenged by the employer. When the amount due under the interim certificate was not paid the contractor issued a writ for the outstanding sum and applied for summary judgment under Order 14 of the Rules of the Supreme Court.

The Court of Appeal had to resolve whether the contract, on its true construction, entitled the contractor to immediate payment and summary judgment under the certificate. It decided that the contractor was not entitled to immediate payment and summary judgment where the employer was able to raise a bona fide arguable dispute as to the correctness of an interim certificate, and that in this case the employer had raised just sufficient material to establish a bona fide arguable contention that the certificate was open to challenge. The court could find no words or any clear or express provision in the contract:

(1)   requiring immediate payment of certified sums where a bona fide dispute as to the correctness of a certificate existed *or*
(2)   making payment of a certified sum a condition precedent to the right to arbitrate.

Similarly, in *R. M. Douglas Construction Ltd* v. *Bass Leisure Ltd* (1990), a case concerning the JCT Standard Form of Management Contract 1987 Edition, it was held that there was nothing within the express terms of the contract to prevent a set-off taking place. The court allowed set-off to take place where there was at least an arguable contention that the payment certificate was incorrect. Furthermore, this included future anticipated claims not yet formally made by the employer against the main contractor.

The same approach was adopted in two other cases: *Re: Clemence Plc* (1992) and *Enco Civil Engineering Ltd* v. *Zeus International Development Ltd* (1991). In *Clemence* an employer failed to pay two interim certificates, even though the works were complete, on the grounds that the works were not of the required standard. The court decided that nothing in the contract (which was similar to the JCT Agreement for Minor Building Works) prevented the company from exercising its common law right to set off damages for breach of warranty against the sums claimed under interim certificates. In *Enco*, which concerned interim certificates of the engineer under the ICE Conditions 5th Edition, the court could find nothing in the conditions to prevent the employer from setting off any cross-claim against the certificates.

Further, in *Rosehaugh Stanhope (Broadgate Phase VI) Plc and Rosehaugh Stanhope (Broadgate Phase VII) Plc v. Redpath Dorman Long Limited* (1990) the Court of Appeal restated the proposition that:

> '... if a contract is to deprive a party of his ordinary right of set off clear words must be used: there is a presumption that neither party intends to abandon any remedies for breach of contract arising by operation of law.'

An example of the stringency of the requirement for clear words to exclude the right of set-off may be found in *Connaught Restaurants Ltd v. Indoor Leisure Ltd* (1993) in which the Court of Appeal found, in the context of a lease, that a provision that rent should be paid 'without any deduction' was not sufficient to exclude the tenant's right of set-off. Similarly, in *BOC Group Plc v. Centeon LLC* (1998) a provision in a contract providing that the obligation to make payment was 'absolute and unconditional' and 'shall not be affected by ... any other matters whatsoever' did not exclude the paying party's right of set-off. (For an example of a wording that was found to effectively exclude set-off, see *Coca Cola Financial Corporation v. Finsat International and Others* (1996)). In *NEI Thompson Limited v. Wimpey Construction UK Limited* (1987) the Court of Appeal, in considering clause 15(3)(d) of the Federation of Civil Engineering Contractors Blue Form of Subcontract, concluded that a requirement for notice in writing of intention to withhold payment to be given before payment became due was insufficient to exclude the contractor's rights of set-off conferred by law.

By innovations introduced in the 1970s, numerous standard forms of subcontract were amended to include express terms restricting the use of the right to set-off otherwise available at law (such exclusions are not prohibited by law, see for example *Coca Cola v. Finsat*). This gave rise to an important distinction between standard forms of main contract and standard forms of subcontract.

However, as a result of the HGCRA a distinction must now be drawn between the position under:

(1) 'construction contracts' – where the right to set-off is restricted by a statutory implied term *and*
(2) 'non-construction contracts' – to which only the common law rules apply.

Subject to the special rules relating to 'construction contracts' it must always be kept in mind that, whatever the position under the general common law, the position concerning set-off will generally be subservient to the express terms of the particular contract between the parties. Therefore, the starting point in considering any question of set-off is always to consider carefully the terms of the contract between the parties.

Later in this book relevant clauses from various standard forms of contract in common use and the provisions of the HGCRA are considered in detail. However, before that it is important that the background law is properly appreciated.

1.2   *The background law*

The crucial question is whether a counterclaim is:

- a full defence
- a counterclaim only
- a counterclaim which is also a set-off which can be utilised as a defence.

All set-offs are counterclaims but not all counterclaims are set-offs. In addition, a set-off which can be used as a defence can be distinguished from a pure defence. The distinction can sometimes be difficult to discern. For instance, if a carpet is purchased based on a sample and when the carpet is delivered the dye of the pattern in a small area of the carpet has run, the purchaser may elect to keep the carpet without prejudice to any claim for breach of warranty. If the purchaser is sued for the price, the inferior quality can be raised as a true defence. The purchaser is saying that the article delivered is not up to sample and is not worth the amount being claimed. This is not a set-off; it is a defence. It derives from the common law right of abatement. As will be seen later when considering the case of *B.W.P. (Architectural) Ltd* v. *Beaver Building Systems Ltd* (1988), this distinction between a set-off and a pure defence is sometimes lost sight of.

The leading modern analysis of the historical context of set-off is contained in the judgment of Morris LJ in *Hanak* v. *Green* (1958), the facts of which have been given at the beginning of this chapter. The issue of set-off became relevant on the question of costs. In the dispute between Mrs Hanak and her builder, the County Court judge gave judgment for Mrs Hanak in the sum of £75 and for the builder under the counterclaim in the total sum of £85. The judge then came to consider the question of costs. He awarded Mrs Hanak her costs on the claim as she had won her claim to the extent of £75. He then awarded costs to the builder on the counterclaim as he had won £85 on the counterclaim.

The builder appealed. He said that in net terms he was the winner by £10 and Mrs Hanak had got nothing. He said that on the question of costs his counterclaim should be treated as a right of set-off and usable as a defence as well as a counterclaim, so he should receive costs but not have to pay any.

In order to determine the proper basis on which costs should be ordered the court had to consider the question of whether the counterclaim could also be used as a set-off which could be used as a defence. So for the sake of a net judgment worth £10 the Court of Appeal delved into the history of set-off. As Morris LJ put it:

> 'So it has come about that we have heard a learned debate, rich in academic interest … on the subject as to whether certain claims could be proudly marshalled as set-off or could only be modestly deployed as counterclaim.'

Held

The court held that the counterclaim was a proper set-off which could be raised as a defence. Lord Justice Morris explained the position as follows:

> '… a set-off when permissable is a defence … The position is, therefore, that since the Judicature Acts there may be (i) a set-off of mutual debts; (ii) in certain cases a setting up of matters of complaint which, if established, reduce or even extinguish the claim, and (iii) reliance upon equitable set-off and reliance as a matter of defence upon matters of equity which formerly might have called for injunction or prohibition.'

Accordingly, Green obtained first his costs of defending Mrs Hanak's claim, and secondly the costs of his successful counterclaim.

1.3   Abatement

Originally there was no common law right at all to set off a counterclaim. The common law did, however, allow the remedy of abatement by which the purchaser could defend himself by showing how much less the subject matter was worth by reason of the breach of contract. The remedy of abatement is explained in *Mondel v. Steel* (1841) and statutory effect has been given to it in relation to the sale of goods by Section 53(1) of the Sale of Goods Act 1979. In the case of *C. A. Duquemin Ltd* v. *Raymond Slater* (1993) it was explained that abatement entitles the purchaser to deduct the difference between the value of the work and materials at the date supplied and their value if they had not been defective. It does not, however, permit anything other than a deduction against the price. The *Duquemin* case, however, left open the question, particularly in the context of building contracts, as to which work and materials are subject to abatement. The point arose for consideration by Judge Lloyd QC, sitting as an Official Referee, in *B.R. Hodgson Ltd* v. *Miller Construction Ltd* (1995).

B.R. Hodgson Ltd v. Miller Construction Ltd (1995) Unreported

The facts

Hodgson was a specialist subcontractor employed by Miller under a DOM/1 subcontract to carry out floor screeding work. The measured work was valued at £81,195.60, of which Miller had paid £50,947.25. Hodgson issued proceedings for the balance due, being £27,317.85. Miller contended that the work was defective because, contrary to the requirements of the subcontract specification, Hodgson had failed to load the slurry grout on the concrete slabs. Hodgson admitted that it had breached the contract but asserted that the cost of the work omitted was £192.08. Miller contended that the cost of the work omitted was £1,358.00. However, Miller also contended that there should be deducted from the price of the works the sum of £32,407.28 by way of abatement of the price because by reason of the breach the floor was valueless. Hodgson argued that the abatement should, at most, be the sum of £1,358.00, representing the value of the materials omitted.

Held

The Official Referee decided that there was nothing in the subsequent authorities materially detracting from or adding to the statement of law in *Mondel* v. *Steel*. Further, he held that the abatement was not limited to the cost of the missing grout. The contract was for the laying of a floor, and as it was contended that the absence of the grout made the floor worthless, Miller should have leave to defend.

By way of example the judge observed that, in a contract for the construction of a simple reinforced concrete beam in breach of which steel reinforcement was omitted, such that the beam was incapable of being safely loaded, the beam for all practical purposes would be wholly valueless. The judge considered that it would be an odd conclusion if the innocent party, in showing how much less the beam was worth, was limited to establishing only the value of the missing steel.

The right to abate can also arise as a result of an express contractual term. In *Barrett Steel Buildings Limited* v. *Amec Construction Limited* (1997) Judge Esyr Lewis QC decided that the words of condition 21.4 of the DOM/1 Conditions, which provide that a claim for interim payment must be for 'the total value of the sub-contract work on site properly executed by the sub-contractor ...', enable the contractor to raise a plea that the work is not 'properly executed', which gives rise to a ground of defence similar to but distinct from the common law defence of abatement.

The right to abate may not, however, be invoked to support delay claims. While some parts of the judgment of Baron Parke in *Mondel* v. *Steel* give support to the proposition that abatement applies to all matters of defence, nonetheless the Court of Appeal in *Mellows Archital Limited* v. *Bell Projects Limited* (1997) contrasted 'depreciation of the value of the work done' with 'consequential damage', the latter of which gives rise to cross-claims which may be deployed by way of equitable set-off rather than abatement.

Mellows Archital Limited v. *Bell Projects Limited* (1997) 87 BLR 26, CA

The Facts

Mellows were subcontactors to Bell under a DOM/1 form of sub-contract. Bell withheld £10,165.49 from the value of the work done because of losses incurred as a result of delay by Mellows. At first instance Judge Wilcox, sitting as an Official Referee, held that Bell's losses could not be asserted by way of set-off owing to the failure to comply with clause 23.2 of DOM/1. Nonetheless the judge held that the claim for delay was an arguable defence to the interim payment claim, if the delay claim could be characterised not as a matter of set-off but as a matter of abatement. Mellows appealed. The Court of Appeal held that the defence of abatement does not include claims that assert losses attributable solely to delay. In particular, as Hobhouse LJ put it:

> 'It is therefore clear that, for a party to be able to rely upon the common law right to abate the price which he pays for goods supplied or work done, he must be able to assert that the breach of contract has directly affected and reduced the actual value of the goods or work – "the thing itself".'

Hobhouse LJ expressly refused to rule out completely the possibility that delay might affect the value of the thing itself; nonetheless he considered that the normal effect of breaches of obligation of timeous performance will be to cause losses to the other contracting party which are consequential on the breach and therefore can only be relied on, if at all, under the principle of equitable set-off.

The right to abate is '... Independent of the doctrine of "equitable set-off" developed by the Court of Chancery to afford similar relief in appropriate cases...' (see the speech of Lord Diplock in *Gilbert-Ash (Northern) Ltd* v. *Modern Engineering (Bristol) Ltd (1974)*). Further, the difference between abatement and set-off:

> '... is only of significance in very particular situations, namely special issues of limitation ... or where, as in our case, a contractual limitation on remedies confines itself to "set-off".'
> (See the speech of Hobhouse LJ in *Mellows* v. *Bell*.)

1.4 *Legal set-off*

Although originally there was no common law right to set off a counterclaim, eventually statute provided a right pursuant to the statutes of set-off (2 Geo II, Chapter 22 and 8 Geo II, Chapter 24) which conferred a right to set-off where the claims on both sides were liquidated debts or money demands which could be ascertained with certainty at the time of pleading. Although statutory in origin, this type of set-off is usually referred to as 'legal' or 'common law' set-off to distinguish it from 'equitable' set-off. The Court of Appeal had occasion to consider the scope of common law set-off in *B. Hargreaves Ltd* v. *Action 2000 Ltd* (1992).

B. Hargreaves Ltd v. *Action 2000 Ltd* (1992) 62 BLR 72, CA

The facts

Hargreaves entered into nine subcontracts on 28 September 1989 with the defendants for the construction of petrol stations. Subsequently Hargreaves went into administrative receivership. The receivers commenced proceedings to recover sums due under one of the subcontracts and applied for summary judgment. The defendant sought unconditional leave to defend on the basis of various legal and equitable set-offs and in support produced the evidence of an independent surveyor who had found that work done under some of the other subcontracts had been omitted or carried out inadequately. The surveyor expressed his opinions on the valuation of omitted and inadequate work and concluded that, in his opinion, there were sums due to the defendant. At first instance, judgment was given for Hargreaves in the sum of £50,000. The defendant appealed on the basis of an entitlement to set-off at common law.

Held

The Court of Appeal held that a set-off at common law is only available where the claims on both sides are in respect of liquidated debts or money demands which could readily and without difficulty be ascertained; and that the claims which the defendant sought to set off were not liquidated debts but were claims which could only be ascertained by litigation or arbitration and were not ascertainable at the date of appointment of the receivers.

1.5   *Insolvency*

Statute also provided a right of set-off in respect of mutual debt provisions on bankruptcy and liquidation. Currently, this right of set-off is to be found in section 323 of the Insolvency Act 1986 (which concerns mutual credit and set-off in bankruptcy) and rule 4.90 of the Insolvency Rules 1986 (which concerns mutual credit and set-off in liquidation).

Section 323 provides:

> '(1) This section applies where before the commencement of the bankruptcy there have been mutual credits, mutual debits or other mutual dealings between the bankrupt and any creditor of the bankrupt proving or claiming to prove for a bankruptcy debt.
>
> (2) An account shall be taken of what is due from each party to the other in respect of the mutual dealings and the sums due from the other.'

Rule 4.90 adopts substantially the same wording and it is therefore unlikely that any distinction exists between the two provisions. The right of set-off provided by section 323 and rule 4.90 must be distinguished from legal and equitable set-off. The right provided by the Act and the Rules is not limited to dealings arising out of contract, although debts must be due between the same parties and in the same right. The mutuality requirement is, however, essential (see *Morris and Others* v. *Agrichemicals and Others* (1995)).

1.6 *Equitable set-off*

The Supreme Court of Judicature Act 1873 section 24(3), now section 49(2) of the Supreme Court Act 1981, introduced the right to raise a counterclaim in the same proceedings. Before this procedural innovation it was necessary to mount a separate action.

Nonetheless, a court of equity could, as a matter of discretion, either:

(1) restrain a party, by injunction or prohibition, from proceeding with an action or the execution of a judgment on a claim pending the outcome of a counterclaim; or

(2) if the counterclaim was particularly closely related to the claim so as to justify it being used as a defence, permit an equitable set-off to stop judgment being given at all if the set-off was valid (see *Dole Dried Fruit & Nut Co* v. *Trustin Kerwood Limited* (1990)).

Following the introduction of the right to pursue counterclaims in the same action a further procedural innovation was introduced by the Supreme Court of Judicature Act 1925, section 41, whereby every matter of equity which might formerly have been relied on to restrain proceedings could instead be relied on by way of defence in those proceedings.

For example, in the case of *Morgan & Sons Ltd* v. *S. Martin & Johnson Co Ltd* (1949) the plaintiff stored vehicles for the defendant and subsequently claimed the agreed rent for so doing. The defendant admitted that the rent was due but one of the vehicles had been stolen and the defendant counterclaimed for the value of the vehicle. The question that arose for determination was whether in the circumstances of the case a court of equity would have recognised the counterclaim as an equitable set-off. It was held that a court of equity would have done so and therefore the counterclaim should be permitted as a set-off and as a defence.

For a more recent and interesting example of the application of equitable set-off see the decision of the Court of Appeal in *Filross Securities Limited* v. *Midgeley* (1998) which, among other things, approved the minority decision of Lord Denning MR in *Henriksens Rederi A/S* v. *THZ Rolimpex; The Brede* (1974) that an equitable set-off is not time barred under the Limitation Acts.

There is some overlap between the equitable doctrine of set-off and the common law doctrine of abatement (see *Gilbert-Ash* v. *Modern Engineering* earlier in this section).

1.7 Summary

In summary, therefore, the position is as follows:

(1) In certain cases a setting up of matters of complaint which, if established, reduce or even extinguish the claim may be raised by way of defence, in that they affect the value of the plaintiff's claim, i.e. there can be an abatement of the price by reference to the difference in value between the subject matter of the contract at the date supplied and the value if the subject matter had not been defective.

(2) There can be a legal set-off pursuant to the statutes of set-off where the claims on both sides are in respect of liquidated debts or money demands which can be readily and without difficulty ascertained.

(3) There can be a set-off of mutual debts in bankruptcy or liquidation. The current statutory provisions enabling the set-off of mutual debts are to be found in the case of individuals in section 323 of the Insolvency Act 1986, and in the case of companies in rule 4.90 of the Insolvency Rules 1986.

(4) There can be equitable set-off which can be used as a matter of defence when the counterclaim is closely connected to the claim.

1.8 Counterclaims within the same contract

While there is no universal rule that claims arising out of the same contract may be set off against one another, this will be permitted either where they give rise to a legal set-off or wherever both claims arise out of, and are inseparably connected with, a single transaction. This latter proposition is derived from the judgment of Lord Hobhouse in the case of *Government of Newfoundland* v. *Newfoundland Railway Company* (1888).

The mere establishment of the existence of some form of counterclaim is insufficient. It must either constitute a legal set-off or be so closely connected with the claimant's demand that it would be manifestly unjust to allow the claimant to enforce payment without taking into account the counterclaim, i.e. it must amount to an equitable set-off. In the Court of Appeal case of *Dole Dried Fruit & Nut Company* v. *Trustin Kerwood Ltd* (1990) Lloyd LJ made the point that:

> 'The claim and cross-claim must arise out of the same contract or transaction, and must also be so unless they are so inseparably connected that the one ought not to be enforced without taking account of the other.'

1.9 Claims arising under different contracts between the same parties

If the same parties have two different contracts it is exceptional (subject to the availability of common law set-off) for a party to be allowed to use a claim under one contract as a set-off in response to a claim made against him on the other. It may amount to a counterclaim but it will not amount to a set-off by way of defence.

Possibly, if the transactions are so closely and directly related that the fact that there is more than one contract is almost accidental, then it is conceivable that a set-off may be used by way of defence. For example, serial contracts under a standing offer might in some circumstances be so regarded. Generally, however, the set-off will be as a pure counterclaim and not as a defence.

Prior to the CPR if the two contracts are so closely linked that it would be unfair and inequitable to allow a claimant to obtain the fruits of his claim until the questions raised by the defendant's counterclaim have been determined, the court may, while allowing judgment to be entered on the claim, nevertheless stay execution of it until the determination of the connected counterclaim. A court could previously do this on the hearing of an application under Order 14 of the Rules of the Supreme Court for summary judgment, although the particular power has not been repeated in the CPR. The court retains its power under Order 47 rule 1 of the Rules of the Supreme Court which provides a discretion to order a stay on such terms as it thinks fit where there are special circumstances which render it inexpedient to allow enforcement of the judgment. However, the connection between the two contracts must be close.

In *Anglian Building Products Ltd* v. *W. & C. French (Construction) Ltd* (1972) Anglian supplied French with prestressed concrete beams under what appears to have been separate supply contracts in respect of three motorways, the M6, M4 and M3. French failed to pay for any of these beams. When Anglian sued for payment in respect of the M6 and M4 beams, French raised a counterclaim alleging that the beams supplied for the M3 were defective. They asked the court to stay execution of the judgment obtained by Anglian in respect of the M6 and M4 until the M3 defective beam claim had been determined. The Court of Appeal refused. Lord Denning, after being satisfied that Anglian would be financially capable of meeting any successful claim in respect of the M3 beams, said:

'... I do not see why this counterclaim on the M3 should be used to hold up payment for the work on the M4 and M6, for which ... French have actually had the money from the employers; they have actually been paid for these very units which have been delivered.'

Accordingly, Anglian were at liberty to enforce their judgment.

Another similar example is to be found in the case of *A. B. Contractors Ltd* v. *Flaherty Brothers Ltd* (1978).

1.10   *Position where there is no contract*

It is not unusual in the construction industry to find that work is carried out by one party at the request of another when terms of contract have either not been considered at all or, if they have, no agreement has been reached on important matters. The result can sometimes be that no contract exists. In such a case the law will often allow a claim by the party complying with the request to be paid a reasonable sum, or what is called a *quantum meruit*. This is based on the value to the party receiving the benefit of the work carried out. Such a claim is based in law on restitution in respect of what might otherwise be regarded as an unjust enrichment to the party receiving the benefit.

It is generally considered that what is relevant is the value of the work to the party receiving the benefit, rather than the cost to the party carrying it out. Clearly, therefore, if the work is in part defective, this will reduce its value to the receiving party, even though it might not reduce the cost of carrying out the work to the other party.

More difficult and, on the present state of the law, an unanswered point, is to what extent a set-off which is not a pure defence will be permitted in respect of a *quantum meruit* claim. This issue was considered but not dealt with, by the Court of Appeal in the case of *Crown House Engineering Ltd* v. *Amec Projects Ltd* (1989) in which Crown House had claimed on a *quantum meruit* basis in respect of mechanical and electrical work which they had carried out for the main contractor, Amec. Amec claimed that, in determining an objective value to them for this work, they could properly reduce the value by having regard to the manner in which, or the time at which, the work was performed. In particular Amec contended that, because of the manner of the performance of the works by Crown House and the fact that they were in delay in completion, Amec had incurred costs including:

(1)   the cost of work which Crown House had agreed to do but had not done and which had to be carried out by others;
(2)   repairs to other contractors' work damaged by Crown House;
(3)   the clearing of rubbish and debris left by Crown House;
(4)   costs associated with commissioning and testing;
(5)   disruption to other subcontractors for which Amec had a contractual liability under the relevant subcontracts.

Amec claimed that all of these could be set off against a *quantum meruit* claim just as if it had been a claim in contract.

As this issue had come before the court on a summary procedure, namely an application under Order 29 of the Rules of the Supreme Court for an interim payment, the court declined to deal with it, stating that it required full and detailed argument. Slade LJ put the question in this way:

> 'On the assessment of a claim for services rendered based on a *quantum meruit*, may it in some circumstances (and, if so, what circumstances) be open to the defendant to assert that the value of such services falls to be reduced because of their tardy performance, or because the unsatisfactory manner of their performance has exposed him to extra expense of claims by third parties?
>
> In my judgment, this question of law is a difficult one, the answer to which is uncertain and may depend on the facts of particular cases'

Accordingly, this important issue, although touched on in *Lachhani v. Destination Canada (UK) Ltd* (1997), is left open for detailed consideration in some future case.

CHAPTER TWO

EXPRESS SET-OFF PROVISIONS AND ADJUDICATION BEFORE THE HOUSING GRANTS, CONSTRUCTION AND REGENERATION ACT 1996, PART II

2.1   JCT forms of main contract

As has already been noted in Chapter 1, prior to the implementation of the HGCRA the JCT forms of main contract did not contain any provision expressly restricting the employer's rights of set-off. As will be observed in Chapter 3, that position has now changed.

2.2   JCT forms of subcontract

Prior to the HGCRA only standard forms of subcontract, such as those published by the Joint Contracts Tribunal, restricted by express provision the contractor's right to set-off against sums otherwise due to subcontractors. As will be seen later, the parties are free to continue to use the versions of those standard forms that predate the implementation of the HGCRA (i.e. the pre-April 1998 versions) and for those contracts which do not constitute 'construction contracts' (see Chapter 3) they may wish to do so. The set-off provisions of the pre-April 1998 versions and the case law relating to them therefore remain relevant in this respect.

2.3   The typical subcontract situation prior to the HGCRA

Standard subcontracts for use under a standard main contract impose obligations on the subcontractor to carry out and complete the works in conformity with the contract (and any contract documents) and the agreed programme. The alleged breach of these obligations often gave rise to a set-off claim by the main contractor.

2.3.1 The position before 1 January 1976

Before 1 January 1976 the 'Green Form' (the Federation of Specialist Subcontractors nominated subcontract for use with JCT 63) and the 'Blue Form' (the non-nominated (domestic) subcontract for use with JCT 63) did not contain direct express provisions dealing with set-off and the contract provisions as a whole had to be considered against the general common law background. A number of cases had come before the courts in relation to set-off in the construction industry. The most notable cases were those of *Dawnays Ltd* v. *F. G. Minter and Trollope & Colls Ltd* (1971), in which the Court of Appeal held that no set-off should be allowed, and the subsequent House of Lords case of *Gilbert-Ash (Northern) Ltd* v. *Modern Engineering (Bristol) Ltd* (1974) which, according to most commentators, effectively overturned the Court of Appeal decision in the *Dawnays* case.

Dawnays Ltd v. *F. G. Minter and Trollope & Colls Ltd* (1971) 1 WLR 1205

The facts

In this case Dawnays were nominated subcontractors for steelwork under the Green Form. Their work was valued in an architect's interim certificate under a JCT 63 main contract at £27,870. The main contractors (Minter and Trollope & Colls jointly) received the money from the employers but refused to pay it to Dawnays, contending that they were entitled to damages for delay caused by Dawnays' defective work.

Held

The Court of Appeal held that, on a true construction of the Green Form, Minter and Trollope & Colls could only deduct sums which were liquidated and ascertained and which were established or admitted as being due. Accordingly, the main contractor was not allowed to set off for alleged defects or delays and had to pay the nominated subcontractor the value of the nominated sub-contractor's work included in an interim certificate.

In the case of *Gilbert-Ash (Northern) Ltd* v. *Modern Engineering (Bristol) Ltd* (1974), which was based on a one-off form of subcontract under a JCT 63 main contract, the House of Lords considered the question of set-off and how it had been dealt with by the Court of Appeal in this case. The tentative conclusion reached was that the Green Form did not, by its express provisions, prevent the main contractor from setting off in those situations where he would otherwise be allowed to do so. In other words, the general law (as outlined in Chapter 1) was not affected by the wording of the contract.

On this basis it is generally contended that the *Dawnays* case was overruled by the *Gilbert-Ash* case. However, there are those who regard the House of Lords' decision as restricted, firstly to the one-off form of subcontract which the House was considering, and secondly to a general statement that the law in relation to set-off applied to the construction industry as it applied elsewhere. This latter statement was the result of indications in Lord Denning's judgment in the *Dawnays* case that interim certificates of the architect were to be treated as equivalent to cash and therefore not subject to claims for set-off. The controversy which remained following the *Gilbert-Ash* case became, so far as JCT subcontracts were concerned, largely academic when, on 1 January 1976, the Joint Contracts Tribunal amended the Green Form. Subsequently the Blue Form was similarly amended. These amendments were reproduced in NSC/4 (and other JCT subcontracts) and DOM/1, both of which provided express contractual machinery to deal with the setting off of counterclaims by main contractors against subcontractors.

2.3.2 The position after 1 January 1976

As a result of the cases on set-off in the construction industry coming before the courts in the early and mid 1970s, and because of the fundamental importance of cash flow both for contractors and subcontractors, express set-off provisions and machinery to deal with this issue were devised and ultimately adopted by the JCT. Those provisions have a fundamental bearing on the cash flow position between main contractor and subcontractor.

If the main contractor wished to set off any claim against sums otherwise due to the subcontractor, the requirements of the contract on set-off had to be strictly complied with or the main contractor would lose the right to set off. He retained the right to pursue his claim but could not avoid paying the subcontractor initially.

Any failure by the main contractor to comply with the set-off requirements was very likely to result in the subcontractor being able to issue a writ, and make an application under the former Order 14 of the Rules of the Supreme Court for summary judgment, with every prospect of success.

In these circumstances it reflected badly on the construction industry of those days that main contractors frequently failed to comply with the set-off requirements. Their salvation in most cases depended on the subcontractor's ignorance of his rights in such a situation which, had he known, would have entitled him to obtain payment in full even where the main contractor had a significant counterclaim against him. The subcontractor's victory might not have been permanent, but would have had obvious cash flow advantages and would have improved his negotiating position when dealing with the counterclaim on its merits.

The set-off provisions of NSC/4 were amended by Amendment No.4 (July 1987), and the other JCT subcontracts and DOM/1 were subsequently made the subject of similar amendments. In 1991 the JCT published a new form of nominated subcontract conditions (NSC/C) to replace NSC/4 and 4a. The wording of the set-off provisions was similar but not identical to that in NSC/4 and 4a as amended in 1987. The case law is largely on the NSC/4 and NSC/4a provisions prior to July 1987. It will remain relevant in relation to those contracts entered into prior to 1 May 1998 (the date the HGCRA came into force), and those contracts entered into after 1 May 1998 where the parties, although utilising a JCT or similar standard form, have not adopted the April 1998 amendments because the agreement does not come within the definition of 'construction contract' (see Chapter 3). While agreements of this type are likely to form part of an ever diminishing group, they may nonetheless continue in use for years to come. In the circumstances the case law on the pre-April 1998 set-off provisions of the JCT and other subcontracts will remain of interest for some time yet despite the advent of the HGCRA.

2.3.3 Set-off under subcontracts before the HGCRA

Prior to the HGCRA the JCT subcontracts NSC/4 and 4a, their successor NSC/C and the subcontracts NAM/SC and IN/SC contained clauses dealing with the rights and machinery of set-off and adjudication in the event that a set-off claim was disputed by the subcontractor.

The DOM/1 provisions relating to set-off and adjudication were in substance the same as those relating to NSC/4.

The position of the contractor

The essential features of the pre HGCRA machinery for NSC/C and its predecessors were:

(1) It contained an express right for the contractor to set off money otherwise due to the subcontractor in respect of, but limited to, any claim for loss and/or expense which had actually been incurred by the contractor by reason of any breach of or failure to observe the provisions of the subcontract by the sub-contractor.

(2) The right to set-off was subject to the contractor being able to satisfy three criteria:

    (a) no set-off relating to delay in completion was possible, unless an architect's certificate had been obtained under clause 12 certifying that the subcontractor had failed to complete the subcontract works within the period or periods provided in the subcontract (not relevant to domestic standard forms of subcontract);

    (b) the amount of the set-off had to be quantified in detail and with reasonable accuracy by the contractor;

    (c) the contractor had to have given to the subcontractor notice in writing of intention to set-off the amount which he had so quantified and had also to state the grounds for the set-off. The notice had to be given not less than 20 days before the money from which the amount was to be set off became due and payable to the subcontractor.

(3) It provided that the rights of the parties in respect of set-off were fully set out in the subcontract and that no other rights whatsoever could be implied as terms of the subcontract relating to set-off.

The position of the subcontractor

If the main contractor satisfied the three criteria, the subcontractor had to follow the procedure to secure the appointment of an adjudicator who could rapidly provide a binding holding decision, subject only to an arbitrator's award or court decision at some later date. The adjudicator was entitled to award part or all of the sum in dispute to the contractor; or to order that the contractor pay it to the subcontractor or to order it to be deposited with a stakeholder; or any combination of these. In *Drake & Scull Engineering Ltd* v. *McLaughlin and Harvey plc* (1992), the court showed itself prepared, in appropriate cases, to grant relief to enforce the award of such an adjudicator. In that case a mandatory injunction was granted to support the contractual machinery.

If the subcontractor failed to comply with the adjudication provisions, the contractor was entitled to retain the sums set off unless and until otherwise ordered by an arbitrator or the court. The most important points for the subcontractor were that he had to:

(1) send to the contractor within 14 days by registered post or recorded delivery a written statement setting out his reasons for disagreeing with the set-off; and at the same time
(2) give notice of arbitration to the contractor; and
(3) request action by the adjudicator and send to him by registered post or recorded delivery a copy of the contractor's notice and his statement in reply.

2.4   A consideration of the cases prior to April 1998

Redpath Dorman Long Ltd v. Tarmac Construction Ltd (1981) 1-CLD-07-32

The facts

This was a case on the Blue Form. Redpath were domestic sub-contractors to Tarmac at the Naval Base at Rosyth. Redpath sought payment of £290,000. Tarmac admitted that this sum had been included in the value certified by the architect. However, they claimed to set off a similar sum as damages for breach of contract due to delay and non-performance. Redpath sought summary judgment. The only quantified items which Tarmac could show they had actually incurred were £76,000 for inefficient utilisation of plant and labour and £3,500 because of difficulty in redeployment of resources. Most of Tarmac's claim related to losses which they would or might incur in the future.

Held

The set-off provisions did not allow any claim for loss and expense which might occur in the future. As all but £80,000 of Tarmac's claim related to expenses to be incurred in the future they could not be set off and accordingly Redpath obtained judgment for £210,000. The £80,000 claim was stayed to arbitration.

Chatbrown Ltd v. *Alfred McAlpine Construction (Southern) Ltd* (1986) 35 BLR 44

The facts

This case is similar in many respects to the Redpath case, and again involved the Blue Form. Chatbrown were domestic subcontractors to McAlpine at an RAF Station at High Wycombe. They provided structural steelwork. The following dates were important:

- July 1984 – contract entered into;
- July 1984 to March 1985 – complaints by McAlpine against Chatbrown regarding delays. By March 1985 Chatbrown had become entitled under the subcontract to almost £232,000 which McAlpine failed to pay;
- March 1985 – McAlpine gave notice of intention to set off enclosing an assessment of costs incurred under the set-off provisions;
- February 1986 – Chatbrown issued a writ claiming £231,840 and sought summary judgment under Order 14 of the Rules of the Supreme Court;
- 13 March 1986 – the date for completion of the main contract works.

In their set-off notice McAlpine had referred to the fact that the completion by Chatbrown of their work was essential for the following trades to begin and that the delay by Chatbrown would have a knock-on effect and delay the overall completion of the work. An assessment was made that the delay would total 16 weeks. In assessing the costs of the 16 week delay McAlpine had included 16 weeks of general site costs amounting to £231,840, which clearly related to a future loss, since in March 1985 there was still a year to run on the main contract.

Held

The judge at first instance held that, as the notice was given many months before the main contract completion date, the preliminary costs to which McAlpine had referred would not actually be incurred until some time after that date. Until the main contractor reached a point at which he would otherwise, but for Chatbrown's delays, have finished on site, the site would remain open and the preliminary costs would be incurred in any event. The claim therefore related to future losses.

McAlpine's argument was that although the money would be laid out at a future date, nevertheless it constituted a present loss or expense at the time when the notice was given. In other words, said McAlpine, the loss and expense had already been incurred. The judge, however, referred to the use of the phrase in the set-off clauses 'loss and/or expense actually incurred'.

In the Court of Appeal, McAlpine made the point that if the judge at first instance was correct in taking this view of when the loss and expense was incurred, the set-off provisions, so far as claims for delay were concerned, would be unworkable at any time prior to the date when the main contract works were due to finish.

The Court of Appeal upheld the judge's decision that McAlpine's notice contained an estimate or future assessment and not a statement of costs which had already been incurred. Kerr LJ observed that, while he was not saying that the set-off provisions could never be used in cases of delay such as the present, there were nevertheless great difficulties in using the set-off provisions in such situations.

Contractors should pay careful attention to the views expressed by the Court of Appeal in this case. It is true that despite delay caused by a subcontractor to the main contractor, expenditure on preliminary items which are time related, e.g. site hutting, security, supervision and so on, will still be incurred in just the same manner as if no such delay had arisen until the contractor's planned completion date is reached. Such expenditure could not form the basis of a set-off claim under the Blue Form, nor under the pre-April 1998 versions of the JCT subcontracts. The position under the April 1998 Amendments is dealt with in Chapter 3 (see section 3.2). There may be some slight differences.

The approach of refusing to admit future loss claims is not wholly satisfactory. The delay may not increase actual expenditure but can nonetheless generate an actual loss. Delay by a subcontractor is likely to delay progress which in turn will result in lower valuations and lower interim certificates than would be the case without the delay. The contractor's reimbursement in respect of preliminary items will therefore be delayed. It is delay in the receipt of money from the employer, rather than an increase in expenditure. Nonetheless, it is a real loss actually incurred. If the contractor is able to put his set-off claim in this way it should in principle satisfy the requirement that the loss or expense should have been actually incurred.

In practice, contractors often seek to use the set-off provisions in the Blue Form and DOM/1 to deduct from sums otherwise due to the subcontractor, the amount of liquidated damages which the main contractor is likely to suffer at a future date when he overruns under the main contract. The use of the set-off machinery in such a manner is open to the same objection as was successfully sustained by Chatbrown. The April 1998 amendments do not appear to have changed the position.

Archital Luxfer Ltd v. A. J. Dunning & Sons (Weyhill) Ltd (1987) 47 BLR 1

The facts

This was a Court of Appeal case dealing with a nominated subcontractor for aluminium windows under NSC/4. Archital were the nominated subcontractors and Dunning the main contractor. Archital were to supply and install aluminium windows at a hostel in Basingstoke. The case concerned whether the three main criteria in clause 23.2 had been satisfied by the main contractor. These were:

(1) clause 12 certificate from the architect in relation to a delay claim;
(2) quantification of the set-off in detail and with reasonable accuracy;
(3) notice in writing expressing an intention to set off the amount quantified at least 20 days before the money became due and payable.

Dunning set off from six architect's interim payment certificates issued between 29 January 1985 and 28 January 1986 totalling £26,516. The main issue before the Court of Appeal was whether a letter from Dunning of 29 January 1985 was sufficient in terms of quantification with reasonable accuracy and whether it expressed the intention to set off; in addition, it was necessary to decide whether the architect's letter of 30 January 1985 amounted to a clause 12 certificate.

The judge at first instance had held that whether such letters satisfied the requirements of clause 23.2 was arguable in relation to five of the certificates, and he therefore had stayed the claim of Archital in respect of these to arbitration. In relation to the first certificate of 29 January – as Dunning's notice was dated 29 January and the architect's alleged certificate was dated 30 January – the last of the three criteria could not have been satisfied in respect of this certificate and the judge had accordingly given summary judgment for £1,541. It was the balance of about £25,000 which was the subject of the Court of Appeal's consideration.

Held

If the architect's letter of 30 January 1985 amounted to a clause 12.2 certificate regarding Archital's delay, and if the Dunning letter of 29 January amounted to both sufficient quantification with reasonable accuracy and an expression of an intention to set off, then in relation to the last five of the six certificates, Dunning would have satisfied the criteria.

In relation to the clause 12.2 certificate, the architect's letter of 30 January 1985 included a paragraph as follows:

'We confirm that Archital ... have caused a delay to the North Block in not completing works in this area in the time originally agreed between yourselves and them.'

It was Archital's contention that the North Block was only part of the subcontract works and did not have a separate period for completion in the subcontract documentation and in particular in the item at paragraph 1.C of Schedule 2 to NSC/1. Dunning's response was that by an exchange of letters the provisions of paragraph 1.C of Schedule 2 had been varied by agreement and that the North Block had its own completion period. Accordingly, Dunning said the architect's letter was a valid certificate in relation to the North Block or at least sufficiently arguably so that the matter was a dispute which should be stayed to arbitration.

The Court of Appeal held that in these circumstances it was at least arguable that Dunning had satisfied the criteria in clause 23.2.1.

Regarding the requirement for quantification in detail with reasonable accuracy, Dunning's letter dated 29 January 1985 included the following words:

> 'We must therefore make a provisional interim claim for all liquidated damages, loss and expense, incurred by this company, as follows.'

There then followed a provisional assessment of 10 weeks delay which listed certain matters, e.g. scaffold, plastering, carpentry, painting, and put lump sum figures against them. It also contained a sum in respect of site offices, overheads and expense, again as lump sums. The letter concluded by reaffirming that the figures were provisional costs and that full details would be made available when the facts were known and substantiated.

Archital argued that this was insufficient in terms of quantification with reasonable accuracy and that far more detailed information and substantiation should be provided. Dunning said that sufficient information was given.

The court held that the letter contained at least arguably sufficient quantification for the purposes of clause 23.2.2. The court did not consider whether it was adequate but simply declared that it was arguably adequate so that Dunning may or may not have satisfied the requirement. As there was a dispute about this it should go to arbitration provided the other criteria in clauses 23.2 had been satisfied.

The court commented that the principal purpose of the procedure was to enable a subcontractor to operate the adjudication provisions in clause 24, including the subcontractor's written statement setting out his reasons for disagreeing with the contractor's notice. Looked at in this light it was arguable that there was sufficient in Dunning's letter of 29 January to enable Archital to answer the contentions sufficiently to enable the adjudicator to fulfil his function.

Regarding the requirement for a notice of intention to set off, the paragraph referred to above from Dunning's letter of 29 January contains no express statement that Dunning intended to set off the amount of their provisional interim claim from monies otherwise due to Archital. The court admitted that it had greater difficulty over Dunning's contentions as to the arguability of compliance with the requirement for notice of intention to set off, than it had with the other criteria. However, the court said that it was clear from

Archital's letter in reply of 31 January that Archital realised what Dunning intended to do and that the letter of 29 January was being relied on by Dunning as justifying the set-off against future certified sums.

Accordingly, Dunning had arguably satisfied all three criteria and could therefore retain the sums set off unless and until an arbitrator awarded otherwise. There would be no summary judgment in respect of these sums in favour of Archital.

B.R. Hodgson Ltd v. *Miller Construction Ltd* (1995) Unreported

The facts

This was a decision of Judge Loyd QC, sitting as an Official Referee. Hodgson were tiling subcontractors under a DOM/1 form incorporating, inter alia, Amendments 1, 2, and 4–9. Hodgson were entitled to £659.92 subject to Miller's entitlement to set off in respect of its claim for £960.90. It was not in dispute that the first payment became due on 31 May 1993 and subsequent interim payments were due at intervals of one month thereafter. Miller gave notices of set-off on 18 January 1994 (£58.20), on 25 August 1994 (£75.00) and on 5 November 1994 (£827.70) amounting in total to £960.90. Hodgson contended that the notices were invalid as they were not given within the time limit described in clause 23.2.2 of DOM/1 which (at that time) provided:

> 'Such notice shall be given not less than 3 days before the date upon which the payment from which the Contractor intends to make the set-off becomes due under Clause 21.2.1 or Clause 21.2.2.'

Hodgson contended that the notices had to be given by the 14th of each month in respect of the payment due in that particular month.

Held

The judge decided that it was arguable in the light of *Archital v. Dunning* that the notices, none of which specified that they applied only to the month in which they were issued, were valid to the extent that they could affect later valuations and therefore gave Miller unconditional leave to defend.

Pillar PG Ltd v. *D. J. Higgins Construction Ltd* (1986) 34 BLR 43

This is another Court of Appeal case. This time the subcontractor was nominated under NSC/4a.

To follow the argument in this case it is important to remember some of the wording in the set-off provisions:

> '23.2 The Contractor shall be entitled to set off ... the amount of any claim for loss and/or expense...;
> 23.2.1 ... no set-off relating to any delay ... shall be made unless ... the certificate of the Architect referred to in clause 12.2 has been issued to the Contractor...;
> 23.2.2 the amount of such set-off has been quantified in detail and with reasonable accuracy by the Contractor;
> 23.2.3 the Contractor has given to the SubContractor notice in writing specifying his intention to set off the amount quantified in accordance with clause 23.2.2...'

The facts

Pillar were nominated subcontractors and Higgins the main contractor. Pillar were entitled to £7,720 and sued for that sum but Higgins raised a cross-claim for £12,000 for disruption. Higgins did not give notice of intention to set off this sum. They claimed that while such a notice was necessary in respect of a claim in respect of delay, it was not necessary in respect of a claim for disruption.

Higgins put their argument that no notice of intention to set off was necessary in respect of disruption claims in the following way:

- Clause 23.2.1 clearly relates to delay.
- Clause 23.2.2 talks about 'such set-off'. That, said Higgins, related back to the reference to set-off in clause 23.2.1 so can only apply to set-off relating to delay as contrasted with disruption. Therefore, when clause 23.2.2 talks about the amount of such set-off having been quantified in detail and with reasonable accuracy, it is referring to the set-off for delay in clause 23.2.1 which has nothing to do with disruption.
- Clause 23.2.3 requires the contractor to give notice to the subcontractor in writing specifying his intention to set off the amount quantified in accordance with clause 23.2.2. In other words again relating to the quantification of the set-off in respect of the delay claim.

Thus, said Higgins, there is no need for a notice of intention to set-off to be given if the set-off is not in respect of delay but is in respect of disruption.

Held

The Court of Appeal held that the reference to 'the amount of set-off' in clause 23.2.2 dealing with the requirement to quantify in detail, relates back to the introductory words in clause 23.2 which provides that the contractor can set off any claim for loss and/or expense by reason of any breach of or failure to observe the provisions of the subcontract by the subcontractor, and that, therefore, the quantification and notice provisions applied also to disruption claims. Accordingly, as the 20 day notice had not been given, there was no right to set off and Pillar were entitled to summary judgment under Order 14.

Precisely the opposite argument was raised in the case of *Hermcrest Plc* v. *G. Percy Trentham Ltd* (1991), namely that the set-off machinery applied only to disruption claims and not to delay claims.

Hermcrest Plc v. G. Percy Trentham Ltd (1991) 53 BLR 104, CA

The facts

This case concerned a subcontract based on DOM/1 including Amendment No.1. It did not therefore include the amended set-off provisions contained in Amendment No.3 of August 1987, even though the DOM/1 in this case was not made until 7 March 1989. Therefore the wording was the same as that in its predecessor, the Blue Form.

The subcontractor, Hermcrest Plc, claimed from the main contractor, G. Percy Trentham Ltd, the sum of £34,460.60, being the amount of an application for payment made by Hermcrest. Percy Trentham claimed to set off from this sum in respect of alleged delays by Hermcrest.

It was the contention of Percy Trentham that, even if they had failed to comply with the appropriate notice provisions in respect of set-off contained in clause 23 of DOM/1, they could nevertheless still set off, as the clause 23 notice provisions, they maintained, did not apply to delay claims but only to disruption claims.

Clause 12 of DOM/1 provided as follows:

'Failure of Sub-Contractor to complete on time
12.1 If the subcontractor fails to complete the subcontract works within the period or periods for completion or any revised period or periods as provided in clause 11, the contractor shall so notify the subcontractor in writing within a reasonable time of the expiry of that period or those periods.
12.2 On receipt of the notice referred to in clause 12.1 the sub-contractor shall pay or allow the contractor a sum equivalent to any loss or damage suffered or incurred by the contractor and caused by the failure of the subcontractor as aforesaid.'

It should be noted that the reference here is to the main contractor claiming in respect of any 'loss or damage'.

By virtue of clause 13 the contractor is able to claim in respect of disruption caused to the progress of the works by the subcontractor. Clause 13.4 states:

'If the regular progress of the Works is materially affected by any act, omission or default of the Sub-Contractor, his servants or agents and if the Contractor shall within a reasonable time of such material effect becoming apparent make written application to the Sub-Contractor, the agreed amount of direct loss and/or expense thereby caused by the contractor may be deducted from any monies due or to become due to the subcontractor or may be recovered from the subcontractor as a debt...'

Note that the reference is to 'loss and/or expense' rather than to loss or damage.

By clause 23.2, dealing with set-off, it is provided that:

'The Contractor shall be entitled to set off against any money (including any Subcontractor's retention) otherwise due under the Subcontract, the amount of any claim for loss and/or expense which has actually been incurred by the Contractor by reason of any breach of, or failure to observe the provisions of the Sub-contract by the Subcontractor...'

Percy Trentham therefore contended that no notice of intention to set off under clause 23 was required in respect of delay claims under clause 12, as this referred to loss or damage, whereas clause 13 dealing with disruption and clause 23.3 both referred to loss and/or expense.

Held

Judge John Newey QC, sitting as an Official Referee, did not accept the contentions of Percy Trentham. He said:

'Under clause 23.2 set-offs may be in respect of any "loss and/or expense incurred by the Contractor" because of breach of contract by the subcontractor. Set-offs may, therefore, arise out of delay in performance, disruption or any other failure to comply with the subcontract.

I think that for a contractor to set off against an interim payment under clause 21 an amount claimed for failure to complete works within the original period or any extended period, he must first have given notification of the delay within clause 12.1 and afterwards quantified the amount in detail and with reasonable accuracy and given notice of intention to set it off 20 days before the interim payment fell due as required by clauses 23.2.1 and 2.

A contractor who has given notice of delay within clause 12.1 but not complied with clauses 23.2.1 and 2 will be able to bring a claim against the subcontractor, but he cannot claim any rights of set-off as clause 23.4 confined the parties to rights set out in the subcontract.'

Accordingly, judgment was given in favour of Hermcrest Plc for the sum of £34,460.60.

Note: In *Pigott Foundations Ltd* v. *Shepherd Construction Ltd* (1993) it was decided that the contractor is not entitled to rely on clause 13.4 in the absence of an agreement as to the amount which is to be set off or recovered as a debt, with the result that until the amount has been agreed there is no means to determine how much the contractor is entitled to claim under the clause. Judge Gilliland QC, sitting as an Official Referee, was unimpressed by the argument that the simple expedient of the subcontractor failing or refusing to agree the amount of direct loss or expense would render clause 13.4 ineffective.

Tubeworkers Ltd v. Tilbury Construction Ltd **(1985) 30 BLR 67**

This was again a decision of the Court of Appeal. This time the subcontractor for structural steelwork was nominated under the Green Form.

The facts

The architect issued a certificate including £55,000 in respect of the work of Tubeworkers. Under the Green Form they became entitled to be paid this sum on 4 October 1983. On 9 October the architect issued a certificate of non-completion by the subcontractors of their subcontract. Thereafter, Tilbury quantified their set-off in the sum of £97,000 and notified Tubeworkers on 16 November. On 21 November Tubeworkers issued a writ for the certified sums and sought summary judgment under Order 14.

At first instance the judge gave Tubeworkers judgment, holding that Tilbury had failed to give the required notice 20 days before the certified sum became due and payable to Tubeworkers. However, the judge decided to stay execution of the judgment pending the determination of Tilbury's counterclaim. Tubeworkers appealed to the Court of Appeal against the stay.

Held

The Court of Appeal held that the cash flow position between the parties was to be determined by reference to the subcontract. Under the subcontract, because Tilbury had failed to give the required notice in time, they had no right to set off, even if they had a right to pursue their counterclaim in the ordinary way. Accordingly, the judge should not have interfered with the contractual machinery between the parties which clearly envisaged that Tubeworkers should be paid the certified sum if Tilbury failed to comply with the requirements of the set-off provisions.

The ordering of a stay would usurp the function of the adjudicator who may be appointed under the subcontract.

The court had jurisdiction under Order 47 to order a stay of execution if there were special circumstances making it expedient to do so. Here there were no such special circumstances and, indeed, the express subcontract machinery dealing with set-off adequately provided for the situation.

B. W. P. (Architectural) Ltd v. Beaver Building Systems Ltd (1988) 42 BLR 86

The facts

This case concerned the NAM/SC form for use by named sub-contractors under the Intermediate Form of Building Contract. The set-off provisions are contained in clause 21 and 22 and are similar to those in DOM/1. The main contractor refused to pay against an interim application for payment. This led to the plaintiff sub-contractor issuing a writ and seeking summary judgment for £45,000.

Held

(1) The main contractor was not entitled to allege that the work to which the application for payment related was not properly executed since this would have the effect of nullifying Clause 21 which contained exclusive machinery by which the contractor could challenge the subcontractor's right to interim payment.

(2) Clause 21.4 limited the right to set-off to the express provisions of clause 21 and excluded the main contractor's common law rights of set-off.

(3) The fact that the main contractor may have a bona fide counterclaim did not provide sufficient reason for a stay of execution of a summary judgment.

Note: The first part of the decision, in paragraph (1) above, was regarded by many as highly doubtful. The set-off machinery only comes into play once a sum would otherwise be due to the sub-contractor apart from set-off claims by the main contractor in relation to delay or disruption. If the challenge of the main contractor is in relation to the value of the work done in the first place, it is submitted that that challenge, so far as NAM/SC (and DOM/1) are concerned, must be available to the main contractor. (See the Court of Appeal cases of *Acsim (Southern) Ltd v. Danish Contracting & Development Co. Ltd* (1989) and *A. Cameron Ltd v. John Mowlem & Co. plc* (1990) as a result of which it is doubtful whether this case would be followed and is most likely impliedly over-ruled.)

Acsim (Southern) Ltd v. Danish Contracting & Development Co. Ltd (1989) 47 BLR 55

The facts

The defendant (Dancon) engaged Acsim under the Blue Form of subcontract subject to certain alterations agreed by the parties. Clause 13 required interim payments at monthly intervals comprising the total value of the subcontract work properly executed. This was subject to a special non-standard requirement as follows:

> 'The subcontractor will be paid instalments. Normally the subcontractor shall send in his invoice on the 27th of each month. He will then be paid by cheque sent by the Contractor no later than the 15th of the following month. The retention money is 5% of the invoiced amount...'

Clause 15 dealing with set-off required the giving of written notice before any set-off could be validly effected against monies otherwise due to the subcontractor Acsim. The relevant facts were as follows:

(1)   On 27 August 1988 Acsim submitted an interim application to Dancon showing a net sum due of £221,018.03 on the basis that Acsim had completed 98% of the subcontract works;
(2)   Payment under the application was due on 15 September 1988;
(3)   On 3 October 1988 Dancon wrote to Acsim setting out 14 items amounting in value to £453,000;
(4)   Acsim issued a writ on 7 October 1988 and applied for summary judgment under Order 14 of the Rules of the Supreme Court on 9 November 1988;
(5)   On the hearing of the application for summary judgment an affidavit was before the court on behalf of Dancon asserting that Acsim had not completed their work, that Dancon was investigating whether Acsim's work had been properly executed and disputed the amount properly due to Acsim under the interim application.

Notwithstanding (5) above the judge would not allow an adjournment to enable Dancon to complete their investigations and entered judgment for Acsim because Dancon had not given notice to set-off in accordance with clause 15. Dancon appealed.

On the appeal it was common ground between the parties that Dancon had not by 15 September 1988 given the required notice of set-off in accordance with clause 15 of the contract and, further, that all of the 14 items in the letter dated 3 October 1988 were matters of set-off. The appeal turned on the question of whether the money claimed under the interim application was 'money otherwise due' under the subcontract. Further affidavits were before the court at the appeal hearing which demonstrated as arguable propositions that Acsim's work was overvalued, not completed and not properly executed. Dancon claimed to be entitled to defend themselves by showing that Acsim had not done all the work in respect of which the claim was made and/or that the amount claimed was wrongly calculated and/or that the work done had been badly done in breach of the subcontract and was worth less than the sum claimed in respect of it. Acsim, on the other hand, claimed that the effect of the alteration to clause 13 was that the amount shown in the interim application dated 27 August 1988 was 'money due' without any right of defence available to Dancon, save for that set out in clause 15.

Held

On a true construction of the terms of the subcontract, Dancon had not lost the right to dispute the amount of an interim application simply by failing to comply precisely with the set-off provisions of clause 15. Dancon could still dispute the value of the certificate by showing that it included the value of works not in fact done; or that it was calculated on a mistaken basis as to the agreed cost of the work done; or that part of the work done was worth less then the agreed price by reason of breach of the subcontract terms.

In reaching its decision the Court of Appeal rejected the findings of the court in the *Beaver* case that the set-off provisions contained the 'exclusive machinery' by which a contractor can challenge a subcontractor's right to an interim payment. This was because there were no clear express words in the contract to show that the party promising to pay had agreed to abandon the right to assert that the subcontractor had not earned the amount claimed or that by virtue of a breach of contract the work was worth less than the sum claimed. In other words the paying party had not given up the right to abate (see section 1.3).

The *Acsim* case clearly shows that the express set-off machinery contained in the various forms of subcontract considered here relates to the situation where, apart from the set-off claim, money is due to the subcontractor in respect of the agreed value of the work done. If the real issue is the correct value in the first place of the work done, then the set-off machinery does not come into operation. The *Beaver* case was, therefore, rightly rejected as it confused the position of set off with that where a pure defence existed. The Acsim decision was subsequently followed in *Barrett Steel* v. *Amec*.

Barrett Steel Buildings Limited v. Amec Construction Limited (1997) 15-CLD-10-07

The facts

By a subcontract dated 27 February 1996, which incorporated the DOM/2 conditions of contract, Amec engaged Barrett to design, supply and erect the structural steelwork for a building at Fazackerly Hospital, Liverpool. Amec failed to pay two interim payments. Barrett issued proceedings and applied for summary judgment under Order 14 of the Rules of the Supreme Court.

Amec served a defence and a counterclaim alleging that, in breach of terms of the subcontract, the steelwork design was defective and the steelwork unfit for its purpose in that it deflected excessively under load with the result that there was excessive deflection in the floor slabs in certain bays. This meant that the finished levels exceeded the expected tolerances and that the floor slabs sagged to an unacceptable degree. As a consequence, Amec alleged that it was necessary to carry out remedial works to rectify the surface level of the floor slabs at a cost of £47,845 plus VAT and that they were consequently entitled to an abatement of the price of the subcontract works.

The abatement was pleaded in the defence as follows:

'Amec are entitled to claim, and claim, a right of abatement at common law, that is any (which is denied) outstanding liability to Barrett is reduced and extinguished by reference to the diminution in value of the sub-contract works by reason of the defects therein (pleaded above)(sic)...'

Barrett submitted that the defence did not in reality raise the defence of abatement but merely attached the abatement label to what is in fact a set-off based on the cross-claim contained in the counterclaim.

The 'total cost of remedial works' was pleaded as £47,845 (plus VAT) but the defence did not expressly allege that the reduction and extinguishment of any liability were to be measured by the cost of remedial works which were the subject of the counterclaim and the set-off.

Held

Judge Esyr Lewis QC rejected Barrett's submissions in these terms:

> 'I do not find anything in the judgment of Ralph Gibson LJ [in *Acsim*] or the other judgments in the *Acsim* case to suggest that, where a contractor claims a set-off under the terms of the clause 23, he is precluded from running an abatement defence or any other defences not in the nature of a set-off which he may have in tandem with a set-off.'

Further, it was concluded that a defendant in a supply contract or a contract for the sale of goods may rely on the cost of remedying defects as a measure of the reduced value of the goods or services provided.

The Court of Appeal decision in *A. Cameron Ltd* v. *John Mowlem and Company Plc* (1990) demonstrated the fact that the contractor could, despite following the set-off procedures through to an adjudication, subsequently refuse to honour the adjudicator's decision by contending (belatedly) that the valuation against which the contractor originally claimed to set off, was itself wrong.

A. Cameron Ltd v. *John Mowlem & Company Plc* (1990) 52 BLR 24, 8-CLD-07-01

The facts

John Mowlem were the main contractor with Cameron as the sub-contractor under the DOM/1 form of subcontract. The DOM/1 form was signed in December 1987 but was based on the form prior to Amendment No.3 of August 1987.

Cameron made usual monthly applications for payment. On 7 March 1989 Mowlem wrote giving notice of intended set-off under clause 23.2.2 in a quantified amount of £52,800. Cameron responded under clause 24 seeking adjudication and providing a statement of disagreement with Mowlem's intention to set-off.

The matter proceeded to adjudication and the adjudicator's decision was that Mowlem should pay to Cameron the sum of £52,800. Mowlem refused to do this and Cameron issued proceedings and claimed payment by way of application for summary judgment under Order 14 of the Rules of the Supreme Court.

It should be noted that clause 21 4 of the relevant DOM/1 states what amounts to the gross monthly valuation for interim payment purposes as, inter alia, including 'the total value of the sub-contract work on the site properly executed by the subcontractor...'. There is no contractual fetter on the contractor raising an argument as to what the correct valuation should be. As Mann LJ observed in giving the Court of Appeal's decision:

> 'The contractor's right to dispute what is the amount properly due is to be contrasted with his right to set off against an amount properly due.'

Mowlem argued that in effect the amount of Cameron's valuation could still be challenged and that such a challenge, if successful, meant that there was no need to set off at all.

Held

The Court of Appeal quoted the proviso contained in clause 24.4.2 of the adjudication provisions in DOM/1:

> 'The Contractor shall not be obliged to pay a sum greater than the amount due from the Contractor under clause 21.3 in respect of which the Contractor has exercised the right of set-off referred to in clause 23.2.'

Mann LJ went on:

> 'The effect of this proviso is that the contractor who is subjected to an obligation to pay by the adjudicator, does not have actually to pay more than would have been due on the interim payment against which the whole or part ... of the set-off was to have had effect ... the Adjudicator has no power ... to determine the amount due in accordance with clause 21.3. Nor has the Court. In the case of disagreement (as here) the correct figure can be determined only by arbitration.'

Cameron's contention that by giving notice of set-off in respect of the sum of £52,000, Mowlem were admitting that at least this much was due (subject to the set-off) by reason of an estoppel by convention, was rejected by the Court of Appeal.

In the result Mowlem were successful in resisting Cameron's claim in respect of the adjudicator's decision of £52,800 (see also *Barrett Steel Buildings Limited* v. *Amec Construction Limited*, earlier in this section)

Note: As a result of *Cameron* v. *Mowlem,* coupled with the *Acsim* case discussed earlier, it is clear that so far as domestic-type standard forms of subcontract are concerned (and this includes NAM/SC), it was relatively easy for the contractor to side-step the set-off or adjudication machinery by contending, whether initially or even after being involved in an unsuccessful adjudication, that the sub-contractor's valuation was itself subject to dispute.

Under the HGCRA an adjudicator may decide any dispute referred to him on terms that his decision is binding until such time as the dispute is decided by legal proceedings, arbitration or agreement. The effect is that if it remains open to the paying party to deploy a challenge to valuation as a means of avoiding payment of an adjudicator's decision, nonetheless such a challenge may itself be promptly referred to adjudication (see Chapter 3 section 3.2.2).

William Cox Ltd v. *Fairclough Building Ltd* (1988) 16 Con LR 7, CA

The facts

This case involved NSC/4. The relevant sequence of events was as follows:

- 12 March 1987 – contractor gave notice of set-off in the sum of £66,945. Of this, £4,564 related to scaffolding charges with the balance being in respect of 11 weeks delay;
- 17 March 1987 – architect issued a payment certificate in the sum of £8,345.06;
- 3 April 1987 – certificate became payable;
- 17 April 1987 – contractor wrote to the subcontractor specifying further costs incurred in the sum of £10,301.67, but the letter did not expressly refer to the intention to set off these monies;

- 4 June 1987 – architect issued a letter specifying failure to complete by the nominated subcontractor;
- 10 June 1987 – architect issued a further payment certificate showing the sum of £5,191.52 as due to the subcontractor;
- 27 June 1987 – the certificate issued on 10 June 1987 became payable.

Three issues arose:

(1) Whether the contractor's letter dated 17 April 1987 constituted a notice of intention to set off;
(2) Whether the architect's letter dated 4 June 1987 amounted to a certificate of delay (the subcontractor had contended that, as the architect had admitted in the letter of 4 June 1987 that he was still considering the position with regard to extensions of time, no certificate under clause 12.2 could be issued);
(3) Whether the certificate of delay to be effective should have been issued prior to the date on which the monies, from which the deduction was to be made, became due.

Held

(1) The contractor's letter, although it did not in terms state an intention to set off the sum referred to, did by implication have this effect.
(2) The architect's letter was a certificate within clause 12.2. If a valid certificate under clause 12.2 could not be granted until the last application for extension of time had been determined, it would be simple for a subcontractor, by constant application for extensions of time, to prevent an effective certificate being issued until a very late stage.
(3) Unless the court were to find that the certificate under clause 12.2 has to be given by the date the monies become due, the subcontractor's position might well be defeated at any moment up to judgment by the issue of a clause 12.2 certificate. Therefore, looking at the provisions of the subcontract as a whole the court was satisfied that where the contractor has to rely on a certificate under clause 12.2, it must have been issued prior to the date on which the monies become due to the subcontractor.

In this case the certificate under clause 12.2 was issued on 4 June

1987, which was after the date on which the interim certificate dated 17 March 1987 became due for payment. Accordingly, the sub-contractor's claim under that certificate succeeded. However, as the certificate under clause 12.2 was issued before the interim certificate dated 10 June 1987 became due for payment, the subcontractor's claim under this interim certificate failed.

Mellowes PPG Ltd v. *Snelling Construction Ltd* (1989) 49 BLR 109

The facts

This case concerned NSC/4, under which Mellowes were sub-contractors and Snelling the main contractor. Snelling claimed that Mellowes were causing delay. The architect granted Snelling an extension of time of 5 weeks under the main contract in respect of delay on the part of Mellowes, and also certified under clause 12 of NSC/4 that Mellowes' works, which had been completed on 17 April 1989, should have been completed some 12 weeks earlier.

In April 1989 the architect issued two valuations in favour of Mellowes, both of which became payable on 11 May 1989 but neither of which were paid by Snelling who had on 25 April 1989 received a letter from the employer notifying his intention to deduct £32,100 by way of liquidated and ascertained damages for 17 weeks delay to the main contract. The employer had acknowledged that Snelling had applied for a further 10 weeks extension on account of delays by Mellowes and confirmed that, if further extensions of time were granted, the liquidated and ascertained damages would be adjusted accordingly.

On 18 July 1989, Snelling gave notice to Mellowes claiming an entitlement to set-off £46,935.16 from the sums certified in April, some 3 months after the date on which, in accordance with clause 23.2 of NSC/4, the notice should have been given. Mellowes issued a writ and sought summary judgment under Order 14 of the Rules of the Supreme Court.

Held

The court held that if the contractor fails to comply with clause 23.2.1 as regards notices, he loses the right to set off altogether and that the contractor's notices were, in this instance, too late to satisfy the strict provisions of clause 23. Snelling therefore resorted to an alternative argument centred on clause 21.3.1.2, which provides that where the employer has exercised any right under the main contract to deduct from monies due to the contractor in respect of some act or default of the subcontractor, the amount of the deduction may in turn be deducted by the contractor from any monies due or to become due to the subcontractor.

Snelling contended and the court accepted that clause 21.3.1.2 allowed a right of deduction independent of clause 23 and, therefore, one that was not subject to the notice requirement. The court accepted that the two contractual provisions were quite independent, with neither governing the other.

It was necessary, therefore to decide whether or not clause 21.3.1.2 had been arguably satisfied. The judge did not think that it had because Snelling were unable to point to any provision in the main contract giving the employer the right to deduct any money from the main contractor in respect of delay to the main contract works caused by a nominated subcontractor. The judge, Mr Recorder Fernyhough QC, emphasised that the opposite was the case because:

> '... if that situation arises then the main contractor is entitled to an extension of time on account of delay by the nominated sub-contractor, which automatically means that the employer may not deduct liquidated and ascertained damages from any sums otherwise due to the main contractor.'

Snelling's further argument, that the contractor could deduct such liquidated and ascertained damages from sums otherwise due to the subcontractor in the interval before the architect assesses whether or not the main contract works have been delayed by the nominated subcontractor, was also rejected. This rejection was arrived at on the basis that no such right could be pointed to in the contract. Further, such a right would be inconsistent with the scheme of the main contract, which provided that the liquidated damages may only be deducted once a certificate of non-completion had been issued by the architect.

So far as nominated subcontractors are concerned, it is rarely that a contractor will be entitled to rely on clause 21.3.1.2. An example might arise in relation to the operation of the main contract indemnity provisions.

CHAPTER THREE

EXPRESS SET-OFF PROVISIONS AND ADJUDICATION UNDER THE HOUSING GRANTS, CONSTRUCTION AND REGENERATION ACT 1996, PART II

3.1 The Housing Grants, Construction and Regeneration Act 1996

3.1.1 Genesis

The Housing Grants, Construction and Regeneration Act 1996, (HGCRA) introduced statutory provisions seeking to regulate the right to withhold payment, and so impacts on the law of set-off. The Act had its genesis in the widely held perception in the construction industry that the law of set-off, combined with limitations on the power of the courts to grant summary judgment or to order interim payment, had become a rogues' charter. Subcontractors were particularly aggrieved at their inability to obtain a speedy remedy. Even in the main contract relationship, contractors often found themselves exposed to the whims of the employer.

The Rules of the Supreme Court formerly provided for summary judgment to be obtained in the High Court in those cases where there is no arguable defence to the claim (for the position under the CPR see Chapter 6). However, in those cases where an arguable entitlement to abate the price or to set off could be demonstrated, the court would not grant summary judgment (see *United Overseas Ltd* v. *Peter Robinson Ltd* (1991) and section 6.2) so that the merits of the underlying claim had to be sent for trial, often resulting in delay and substantial cost. At the subsequent trial the abatement or set-off might fail. Nonetheless, the plaintiff would have been deprived of payment pending the trial, giving a cash flow advantage to the party withholding payment who might use that advantage to negotiate a settlement involving payment of a sum less than that properly due. In extreme cases the unpaid party might, as a result of the set-off, become insolvent before the conclusion of legal proceedings.

As explained in Chapter 2, the JCT, in response to the decision in *Gilbert-Ash*, developed express contractual terms restricting the rights of set-off in subcontracts and providing a contractual adjudication procedure enabling a notice of set-off to be referred to an independent third party for a temporary decision as to who should have the disputed sum pending final resolution by arbitration.

Anecdotal evidence suggests that contractors avoided the rigours of the JCT provisions restricting set-off, either by insisting on their deletion (and thereby restoring the common law position) or, in some instances, by inserting alternative provisions conferring greater rights of set-off on the contractor. The doctrine of freedom of contract effectively allowed the party with the stronger bargaining position to dictate the terms of contract, often to the detriment of the weaker party. Further, if the requirements of the JCT set-off provisions (where they applied) were rigorously adhered to, their effect could be minimised. The benefit of contractual adjudication, as originally developed by the JCT, was eroded by the decision in *A. Cameron Limited* v. *John Mowlem & Co plc* (1990) (see section 2.4) which held that an adjudicator was not empowered to determine the sum due under the contract, thereby enabling the contractor to avoid the set-off procedures by disputing the value of the works.

Following widespread dissatisfaction and intense lobbying of Government, a review of procurement and contractual arrangements was set up by Government under Sir Michael Latham. In his final report, *Constructing the Team* (published July 1994) he recommended that provisions should be treated as unfair or invalid (see Recommendation 25) if they enabled the paying party:

'(1)   to seek to exercise any right of set-off or contra charge without:

(a)   giving notification in advance;
(b)   specifying the exact reason for deducting the set off;

(2)   to set off in respect of any contract other than the one in progress.'

The Latham Report was followed by the passing of the HGCRA which established an obligation to give notice before withholding payment in respect of money due under a 'construction contract'.

3.1.2 Scope of the HGCRA

The HGCRA only applies to 'construction contracts' (see section 104), these being contracts:

(1) for the carrying out of construction operations;
(2) arranging for the carrying out of construction operations by others;
(3) providing labour or the labour of others for the carrying out of construction operations.

References to construction contracts include agreements to carry out architectural design or surveying work and to provide advice on building, engineering, decoration or landscaping in relation to construction operations

Construction operations are themselves defined in section 105 of the Act and may broadly be summarised as work or work and materials in connection with building and engineering. Agreements purely for the sale of goods do not come within the definition. Certain operations are expressly excluded from the definition of construction operations (section 105(2)) including, by virtue of the Construction Contracts Exclusion Order 1998 (SI 1998 No. 648), certain agreements arising under statute and under the private finance initiative. The exclusion order also excludes certain finance and development agreements from the ambit of the HGCRA. Further the HGCRA does not apply to contracts with a 'residential occupier' of a dwelling or flat (see section 106). Last, the HGCRA only applies to agreements in writing (defined in section 107).

In relation to construction contracts it is provided by the HGCRA that:

- a party shall have the right to refer disputes to adjudication under a procedure complying with section 108; and
- a party shall be entitled to stage payments unless the duration of the work is or is expressed to be less than 45 days (section 109); and
- the contract shall provide an adequate mechanism for determining what payments become due, when and the final date for payment (section 110); and
- the right to withhold payment is restricted to those circumstances where an 'effective notice' is given (section 111); and
- there shall be a right to suspend performance for non-payment (section 112); and

- pay when paid provisions, except those linked to insolvency, are prohibited (section 113).

Further, where a construction contract does not comply with any of these requirements, the relevant provisions of the Scheme for Construction Contracts will apply as if the Scheme were an implied term of the contract (section 114(4)).

The JCT and other bodies have amended their standard forms of contract with the intention that they should comply with the HGCRA. Despite this, there is nothing to stop the parties using earlier versions of JCT contracts if they wish. To the extent that those earlier versions do not comply with the HGCRA, the consequent effect will be that the relevant parts of the Scheme will be implied. In addition, in relation to those contracts which do not come within the definition of 'construction contract', the parties are free to adopt terms which do not conform with the HGCRA. For these reasons it cannot be assumed that the pre-existing versions of the standard forms of building contract in use prior to the HGCRA will entirely fall into disuse, or that the decisions relevant to them will become entirely redundant.

Where a party wishes to rely on any of the provisions of the Scheme for Construction Contracts it will have to go through the process of establishing that the contract in question:

- is a construction contract; and
- does not relate to a residential occupier; and
- is in writing; and
- does not conform with a particular requirement of the HGCRA.

Adjudicators appointed under the Scheme may find themselves called on to decide on the above matters.

3.1.3 Payment provisions of the HGCRA

The HGCRA provides an elaborate mechanism for determining the amount and timing of payment:

- a party to a construction contract is (except where the work duration is, or is estimated to be, less than 45 days) entitled to payment by instalments, stage payments or other periodic payments (section 109(1));

- construction contracts must provide an adequate mechanism for determining what payments become due and when (section 110 (1)(a));
- every construction contract shall 'provide for a final date for payment in relation to any sum which becomes due' (section 110(1)(b));
- once a sum has become 'due', the paying party must give notice not later than 5 days afterwards 'specifying the amount (if any) of the payment made or proposed to be made, and the basis on which that amount was calculated' (section 110(2));
- the withholding of payment 'after the final date for payment of a sum due' is forbidden unless an effective notice has been given (section 111).

The HGCRA gives the parties some latitude to decide the amounts, the intervals or the circumstances in which payments become due (section 109(2)), the timing of payment notices and of notice to withhold payment (sections 110(2) and 111), and the fixing of the final payment date (section 110). However, if the parties fail to provide adequately for such matters, the relevant provisions of the Scheme apply.

No express sanction attaches to failure to give the payment notice required by section 110. Ultimately this may diminish the importance of the payment notice rendering it perhaps toothless and irrelevant.

There is no requirement in the HGCRA on the form of the payment notice or on the form of the calculation of the amount said to be due. So, for example, despatching a cheque within the relevant period together with the traditional sub-contractor payment pro forma (showing the gross valuation and deductions for retention, discount and previous payments) ought, arguably, to suffice.

The tenor of section 110(2) is that the payment notice may specify an intention to pay less than the sum determined as due. This is understandable if it allows the paying party to give effect to any contractual entitlement to discount, retention, previous payments or any other permitted deductions (such as a right arising under a pay when paid provision). Whether sections 110(2)(a) and (b) have any wider significance, particularly in relation to the right to abate or set off, is open to question (see 'Withholding Payment under the Housing Grants, Construction and Regeneration Act 1996' (1998) 14 Construction Law Journal 180).

3.1.4 Withholding payment

A statutory restriction on the right to withhold payment is conferred by section 111(1) which provides:

'(1) A party to a construction contract may not withhold payment after the final date for payment of a sum due under the contract unless he has given an effective notice of intention to withhold payment.

The notice mentioned in section 110(2) may suffice as a notice of intention to withhold payment if it complies with the requirements of this section.

(2) To be effective such a notice must specify:

(a) the amount proposed to be withheld and the ground for withholding payment, or
(b) if there is more than one ground, each ground and the amount attributable to it,

and must be given not later than the prescribed period before the final date for payment.'

Prima facie, section 111(1) does not restrict the right to abate the price (i.e. the value of the measured work) because:

(1) it does not expressly do so; and
(2) the words of section 111(1), i.e. '... of a sum due under the contract...', arguably allow the paying party the right to contend that there is no sum 'due' because of either defective, incomplete or non-conforming work. Support for this proposition may be discerned, by analogy, from the decisions in *Acsim* v. *Danish*, *Cameron* v. *Mowlem* and *Barrett Steel Buildings Ltd* v. *Amec Construction Ltd* (see – Chapter 2). However, there may be an exception to this approach (see section 3.2.2).

Assuming, as is likely, that the courts interpret section 111(1) as leaving the common law right to abate intact, the paying party under a construction contract will, in principle, be entitled at any time to exercise the right to abate the price without giving any advance notice of the intention to do so. Therefore if defective, incomplete or non-conforming work is discovered after the payment notice under section 110(2) is given, and after the last date for giving notice of intention to withhold payment has passed, the paying party may reduce the amount to be paid by reference to the value of the abatement of the price. Although it will remain open to the paying party to dispute the sum 'due' (as occurred in *Cameron* v. *Mowlem*) the receiving party could overcome this by asking the arbitrator to decide the amount due, the date when it was due and whether an effective notice was given in accordance with the contract.

In principle, therefore, it would appear that the effective notice under section 111(1) will only apply to the withholding of payment by way of set-off. If so, then in order to determine what may be withheld it will be necessary to look at the background law (see Chapter 1), in which case a party wishing to withhold payment will be required to establish that the basis for withholding payment is not prohibited by the contract and is one of the three recognised categories of set-off, and if unable to do so the party withholding payment will be ordered to pay the sum due.

If the preceding propositions are correct, the HGCRA is procedural in nature, at least in relation to set-off, and does not alter the substantive law concerning abatement or set-off. Furthermore, it means that the HGCRA has not addressed the Latham recommendation that set-offs should not relate to any contract other than the one in progress, or the criticism of set-off articulated by Leggatt LJ in *Axel Johnson* v. *MG Mineral* (see Chapter 1).

3.2   *The HGCRA and adjudication*

Under the HGCRA a party to a construction contract has the right to refer disputes for adjudication in accordance with a procedure complying with the HGCRA. The parties are free to develop their own procedure (the JCT has done so for its family of contracts), but if they do not or if it does not comply with the HGCRA, the procedure in the Scheme for Construction Contracts will be implied into the construction contract.

Adjudication under the HGCRA is intended to be quick and cheap. The Act requires that construction contracts shall:

- enable the reference of disputes to adjudication;
- provide for the appointment of an adjudicator within 7 days of the notice of dispute;

- require the adjudicator to reach a decision within 28 days;
- allow the adjudicator to extend the period of 28 days by 14 days with consent of the applicant;
- require the adjudicator to act impartially;
- allow the adjudicator to take the initiative;
- make the adjudicator's decision final and binding until the dispute is finally determined by arbitration, court proceedings or by agreement.

Under the Scheme for Construction Contracts any party to a construction contract may give written notice to every other party to refer any dispute under the contract to adjudication. The notice must set out:

- the nature and brief description of the dispute and the parties involved;
- details of where and when the dispute has arisen;
- the nature of the redress sought;
- the names and addresses of the parties to the contract.

Subject to any agreement between the parties the adjudicator shall be either:

- the person so named in the contract; or
- a person selected by the nominating body named in the contract; or
- where appropriate, a person selected by an adjudicator nominating body.

It is likely that most disputes will, at least initially, be referred to adjudication. Adjudicators may be asked to decide relatively simple issues relating to the withholding of payment, such as whether the section 111(1) notice was given by the prescribed time and whether it sets out the amounts to be withheld and the grounds for doing so. Plainly, if the adjudicator should decide that the notice was late or that it is otherwise not 'effective', he ought to order that the sum due be paid. This procedural approach avoids the need for any investigation of the merits of the underlying set-off.

Adjudicators may find disputes referred to them that raise the question of whether a particular ground referred to in an effective notice constitutes, in law, a set-off. If the adjudicator decides that the ground does not constitute a set-off, he can order payment to be made without any need to investigate the merits of the ground relied on. Indeed, if the cross-claim arises from an unrelated contract which in itself is not a construction contract, the adjudicator may not have jurisdiction to decide on the merits of the cross-claim (see the next section). It will be important, therefore, for claimants and their advisors, when framing the notice of dispute, to confine it to the narrow question of whether the ground gives rise to a set-off where they believe that will be sufficient to resolve matters. This approach should keep the costs of preparation and the adjudicator's fees at a modest level. If the adjudicator decides that the ground does give rise to a set-off, a fresh adjudication may be commenced, in appropriate cases, challenging the merits of the ground or grounds for set-off relied on.

Adjudicators may, whether at the outset or subsequently, be asked to deal with the more difficult issues relating to the merits of an abatement or set-off. Whether adjudication is a suitable forum for deciding, in a period of 28 days, the merits of the underlying dispute is open to question.

Other difficult questions arise. Firstly, how should an adjudicator deal with abatements and set-offs utilised in response to the claim made in the adjudication, and secondly, can abatements and set-offs be deployed in response to an adjudicator's decision as a means of avoiding payment?

3.2.1 Abatement and set-off in response to the claim

Under Part I of the Scheme, by paragraph 20, the Adjudicator '... shall decide the matters in dispute' and 'may take into account any other matters ... which are matters under the contract which he considers are necessarily connected with the dispute'.

If adjudication is commenced in relation to a claim for the value of work done, the employer might respond by alleging that the work is defective. If so, paragraph 20 of the Scheme permits a defects cross-claim to be utilised as a defence against a valuation claim. The defects entitle the paying party to invoke the defence of abatement (see section 1.3) which is necessarily connected with the dispute. As explained in Chapter 2 (see *Ascim* v. *Danish* and *Cameron* v. *Mowlem* and *Barrett* v. *Amec*) it was not necessary to give notice of set-off under the pre-April 1998 contracts in order to exercise the right to abate.

The position is likely to be different, however, where defects are relied on by way of an equitable set-off rather than abatement. In those circumstances the adjudicator should not entertain the defects cross-claim unless it has been the subject of an effective notice (see *Mellows* v. *Bell* – section 1.3).

Allowing a cross-claim into an adjudication would beg the question of whether it can stand as a counterclaim as well as a defence. This is difficult but it is possible that, so far as the cross-claim overtops the original claim, it may constitute a separate dispute so that the adjudicator ought not to make a decision in favour of the counterclaiming party other than by way of dismissing the original claim to the extent that it is extinguished by the cross-claim.

3.2.2 The adjudicator's decision – abatement and set-off in response

The Scheme adjudicator is empowered by Rule 20 of the Scheme to decide the matters in dispute. In so doing he may decide that any of the parties to the dispute is liable to make a payment under the contract, and decide when the payment is due and the final date for payment. This power is subject to any effective notice of withholding of payment.

Further disputes may arise between the parties after the adjudicator has given his decision on a particular dispute. For example, the adjudicator may order that a payment be made but the paying party may discover defects (or further defects) after the adjudicator has given his decision and, as a result, refuse to pay the amount decided on in the decision. The paying party could argue that he is merely abating (this would be the position where he is setting off a defects cross-claim against a valuation claim) or legitimately setting off within the contract (setting off may, however, be difficult because under the Scheme a notice of intention to withhold payment under section 111 will nearly always need to be served before the decision is reached, if it is to be effective in avoiding having to honour the decision itself). It is arguable however that the paying party, if he can establish either an abatement or (subject to an effective notice) a set-off, may be able to resist payment of an adjudicator's award.

There is, however, a further difficulty to be overcome when deploying an abatement or set-off as a ground for avoiding payment of an adjudicator's decision. Namely the effect of Rule 23 of the Scheme which provides:

'23. (1) In his decision, the adjudicator may, if he thinks fit, order any of the parties to comply peremptorily with his decision or any part of it.

(2) The decision of the adjudicator shall be binding on the parties and they shall comply with it until the dispute is finally determined by legal proceedings, by arbitration (if the contract provides for arbitration or the parties otherwise agree to arbitration) or by agreement between the parties'.

By virtue of Rule 23 the parties are bound to comply with the decision until the dispute is finally determined. It is difficult to reconcile the concept of abating or setting off against decisions with the strict wording of Rule 23(2). It is arguable that adjudicators' decisions are not susceptible to abatement or set-off. If so an exception may exist to the general rule, mentioned in section 3.1.4, that the paying party has the right to contend that there is no sum due. It is arguable that Rule 23(2), properly construed, does not prevent the abatement of an adjudicator's decision on the basis that the binding nature of the decision is only as to the amount payable subject to the terms of the contract relating to and affecting payment. However this view may prove inconsistent with the decision in *Macob Civil Engineering Ltd* v. *Morrison Construction Ltd* (1999), which found that decisions are binding even when the adjudicator errs on the facts, law or procedure. It may be that Rule 23(2) is *ultra vires* (see section 3.2.3).

Enforcement of the adjudicator's decision may be sought through the court either by an application under Part 24 of the CPR for summary judgment (see Chapter 6), or exceptionally by way of mandatory injunction or an order for specific performance (see *Macob* v. *Morrison*). It is anticipated that Part 24 CPR will probably be the more favoured route for enforcement of decisions requiring the payment of money. The basis of the application will be that the adjudicator has decided that the sum is due and that any defence to payment 'has no real prospect' of succeeding until the matter is finally determined by the court or by agreement.

The defendant might seek to raise an arguable defence to the claim by relying on a post decision abatement or set-off and arguing that the obligation to pay has been discharged by virtue of the abatement which, in effect, constitutes a form of compliance. Post decision abatement or set-off is not expressly inhibited by Rule 20 or Rule 23(2). Alternatively it may be argued that the underlying basis of the abatement or set-off justifies the court granting a stay of execution of judgment pending the determination of the abatement or set-off as a special circumstance under Order 47 of the Rules of the Supreme Court (expressly preserved by the CPR) or for making a conditional order under Part 24 CPR.

It is uncertain whether the first of the preceding arguments will succeed. If it does it will be seen as seriously undermining Scheme adjudication, especially as the post-decision abatement or set-off will not finally determine the dispute giving rise to the decision. There is some authority indicating that the second argument will not generally find favour. In the case of *B.W.P.* v. *Beaver* (see Chapter 2) it was held that the existence of a bona fide counterclaim did not provide sufficient reason for a stay of execution. Further, in *Tube-workers Ltd* v. *Tilbury Construction Ltd* (Chapter 2) the court, in connection with the old Green Form of subcontract, was unwilling to usurp the function of the adjudicator by granting a stay of execution of a judgment. In that case the court accepted that there might be exceptional circumstances to justify the grant of a stay but that the relevant terms of the contract 'must be paramount and take precedence over the court's discretion under Order 47'.

Whether the terms of Rule 23(2) will be found to override the right to abate or the court's residuary discretion to grant a stay in exceptional circumstances remains to be seen. However, an exceptional circumstance might be found to exist where, for instance, there are competing adjudication decisions as a result of the post decision abatement or set-off being referred to adjudication. If one of the parties insisted on payment of the decision in his favour without taking account of the decision in favour of the other party that might, assuming the right to abate is found to have been abrogated by Rule 23(2), constitute an exceptional circumstance justifying a stay of execution or a ground for making a conditional order.

3.2.3 The adjudicator's decision – effect

The decision of an adjudicator is binding both under the HGCRA and the Scheme until finally determined by legal proceedings, arbitration or agreement. As discussed in section 3.2.2, Rule 23(2) of the Scheme arguably prevents the avoidance of payment of money due pursuant to an adjudicator's decision merely by adopting a new abatement or set-off. If it does then the Scheme (being secondary legislation) may have achieved a result which the HGCRA (being primary legislation) did not. This may mean that Rule 23(2) may be *ultra vires* for that reason alone. In this respect it should be noted that section 108 of the HGCRA requires construction contracts to 'provide that the decision of the adjudicator is binding...'. Further, section 114 of the HGCRA empowers the making of regulations '... about the matters referred to in the preceding provisions...'. The Scheme, in Rule 23(2), provides that an adjudicator's decision '... shall be binding on the parties, ... and they shall comply with it until the dispute is finally determined...'. If the words '... and they shall comply with it...' are *ultra vires* then there may be greater scope for the court to refuse to enforce an adjudicator's decision by reason of a subsequently deployed abatement (or a set-off where an effective notice is given), although this will require the word 'binding' to be narrowly construed.

Interestingly, the JCT adjudication provisions introduced in the April 1998 amendments also require the parties to comply with the decision of an adjudicator. In JCT 98 it is provided that:

'41A.7.1   The decision of the Adjudicator shall be binding on the Parties until the dispute or difference is finally determined by arbitration or by legal proceedings or by an agreement in writing between the Parties made after the decision of the Adjudicator has been given.

41A.7.2   The Parties shall, without prejudice to their other rights under the Contract, comply with the decisions of the Adjudicator; and the Employer and the Contractor shall ensure that the decisions of the Adjudicator are given effect.

41A.7.3   If either Party does not comply with the decision of the Adjudicator the other Party shall be entitled to take legal proceedings to secure such compliance pending any final determination of the referred dispute or difference pursuant to clause 41A.7.1.'

As the requirement that the parties 'comply' is contractual, rather than arising under the Scheme, it does not admit of an *ultra vires* challenge. However, the words 'without prejudice to their other rights under the Contract...' raise the question as to what those other rights may be. They are likely to include the matters referred to in clause 41A.7.1, i.e. the right to arbitrate, litigate or agree. But they probably go further than this and include the rights of set-off and abatement arising under the contract. The JCT and its advisors have therefore drawn a distinction between 'binding' and 'comply' in relation to the effect of adjudicators' decisions.

Macob v. *Morrison* indicates that no ground for challenging a decision occurs where the decision fails to give effect to a defence of abatement or to a set-off. Whether this has occurred may be established by asking the adjudicator to give reasons under Rule 22 of the Scheme. Where it is apparent from the reasons that the adjudicator has decided to order a payment and in so doing has wrongly rejected a defence of abatement or a set-off, this may constitute a ground under Part 24 CPR for making a conditional order or special circumstances under Order 47 (as preserved by the CPR) to stay execution, especially as Rule 12 of the Scheme requires the adjudicator to proceed in accordance with the contract and the law.

Further, if an adjudicator is asked to decide that a ground does constitute a valid set-off, and decides wrongly consequently dismissing the claim, the claimant may (subject to any agreement to refer disputes to arbitration) refer the same claim to the court. If the court is satisfied that the sum claimed is due and that there is no real prospect of the defence succeeding, the court may grant summary judgment under Part 24 CPR (or in appropriate cases judgment on an admission) despite the fact that the adjudicator has issued a decision dismissing the claim. It must be observed in this respect that a judgment under Part 24, once given, is a final judgment on the claim and consequently will amount to a final determination of a dispute as to whether or not the ground relied on in an effective notice constitutes a set-off in law. This is a further reason for ensuring that the notice of dispute by which an adjudication is commenced is carefully framed so as to admit the possibility of the decision being open to challenge on, or by way of, a subsequent application to the court under Part 24 of the CPR.

Finally, an adjudicator's decision may be expressed in peremptory terms. By virtue of Rule 24 of the Scheme, section 42 of the Arbitration Act 1996 applies to Scheme adjudications in the following terms:

'42. (1) Unless otherwise agreed by the parties, the court may make an order requiring a party to comply with a peremptory order made by the tribunal.

(2) An application for an order under this section may be made:

(a) by the adjudicator (upon notice to the parties)

(b) by a party to the adjudication with the permission

of the adjudicator (and upon notice to the other parties), or ...

(4)   No order shall be made under this section unless the court is satisfied that the person to whom the tribunal's order was directed has failed to comply with it within the time prescribed in the order or, if no time was prescribed, within a reasonable time.'

This provision is concerned with the means of enforcement rather than whether or not decisions should be enforced. In any event the power conferred on the court is discretionary (as confirmed in *Macob* v. *Morrison*).

3.3   The JCT standard forms of contract

3.3.1   General

As explained in Chapter 1, the JCT forms of main contract, prior to April 1998, did not expressly restrict the employer's right to exercise the rights of set-off conferred by law. Other forms of main contract and the terms of engagement of consultants adopted a similar approach. This exposed the contractor and consultants (see *Hutchinson* v. *Harris* (1978)) to the risk, even in relation to those sums certified as due by an independent certifier, that the employer might before payment assert an entitlement to withhold payment because of an alleged abatement or set-off. In those circumstances all the contractor could do was to arbitrate or litigate which generally involved requiring the court or an arbitrator to decide the substantive dispute, with all the cost and delay necessarily involved save where a summary remedy, such as judgment under the former Order 14 of the Rules of the Supreme Court, was available. The position has now changed substantially. The HGCRA does not discriminate between main and subcontractors; rather it applies to all agreements that fall within the definition of 'construction contracts'. As a result, main forms of building contract and the terms of engagement of consultants are for the first time subject to a notice restriction affecting the paying party's right to set off.

3.3.2 The JCT amendments to main contracts and subcontracts

The JCT main contracts and subcontracts have all been made the subject of further amendments, issued in April 1998, to ensure that they comply with the HGCRA. The JCT embarked, at the end of 1998, on re-issuing all its family of contracts, consolidating all amendments to date and making some further amendments. Extracts from some of these new consolidated standard forms are given in Appendix 7. The amendments in relation to set-off are in substance the same for all the JCT contracts. This text therefore concentrates primarily on the position under JCT 98 and under NSC/C as amended by Amendment 7 issued in April 1998. The comments here will apply equally to most of the JCT contracts in relation to the issues of abatement and set-off; where the position is different for a specific contract, this is mentioned.

3.3.3 JCT 98

Clause 30 of JCT 98 provides, in part, as follows:

'30.1.1 1. The Architect shall from time to time as provided in clause 30 issue Interim Certificates stating the amount due to the Contractor from the Employer specifying to what the amount relates and the basis on which that amount was calculated; and the final date for payment pursuant to an Interim Certificate shall be 14 days from the date of issue of each Interim Certificate ...

30.1.1 2. Notwithstanding the fiduciary interest of the Employer in the Retention as stated in clause 30.5.1 the Employer is entitled to exercise any right under this Contract of withholding and/or deduction from monies due or to become due to the Contractor against any amount so due under an Interim Certificate whether or nor any Retention is included in that Interim Certificate by the operation of clause 30.4. Such withholding and/or deduction is subject to the restriction in clause 35.13.5.3.2.

30.1.1 3. Not later than five days after the date of issue of an Interim Certificate the Employer shall give a written notice to the Contractor which shall, in respect of the

> amount stated as due in that Interim Certificate, specify the amount of the payment proposed to be made, to what the amount of the payment relates and the basis on which that amount is calculated.
>
> 30.1.1 4 Not later than five days before the final date for payment of the amount due pursuant to clause 30.1.1.1 the Employer may give a written notice to the Contractor which shall specify any amount proposed to be withheld and/or deducted from that due amount, the ground or grounds for such withholding and/or deduction and the amount of withholding and/or deduction and the amount of withholding and/or deduction attributable to each ground.
>
> 30.1.1 5. Where the Employer does not give any written notice pursuant to clause 30.1.1.3 and/or clause 30.1.1.4 the Employer shall pay the Contractor the amount due pursuant to clause 30.1.1.1.'

Certificates remain the basis, and a pre-condition, of payment under JCT 98 (see *Lubenham Fidelities and Investments Co Ltd* v. *South Pembrokeshire District Council and Another* (1986)). The obligation of the architect is to certify the amount 'due'. The words of clause 30.1.1.1 and 30.1.1.2 probably do not inhibit the employer's right without prior notice to dispute the amount certified as 'due' and consequently to abate (see section 1.3) the sum certified where defective, incomplete, non-conforming or over-valued work (or over-valued loss and expense) is the basis of the abatement. Support for this proposition may be drawn, by analogy, from the decisions in *Acsim* v. *Danish*, *Cameron* v. *Mowlem* and *Barrett Steel* v. *Amec* (Chapter 2). If correct, the principle stated in *C. M. Pillings Ltd* v. *Kent Investments Ltd* (Chapter 1) would appear unaffected, at least in relation to abatement. The right of abatement is therefore unaffected and remains unfettered unless either section 110(2) of the HGCRA is construed as imposing a fetter, or clause 30.1.1.4 is construed as overriding the right to abate by making the notice under clause 30.1.1.4 the exclusive machinery for challenging the amount certified by the architect under clause 30.1.1.1 (a similar approach was adopted in *B.W.P.* v. *Beaver* (Chapter 2) for the pre-April 1998 version of NAM/SC but did not ultimately fare well).

The set-off provision in clause 30 is far less elaborately worded than the pre-1998 versions of the JCT subcontracts. The paying party's right to withhold or deduct is confined by the requirement that a notice be given, in the absence of which the right will be lost in relation to the interim certificate in question. Clause 30.8 provides a virtually identical arrangement in relation to the final certificate.

Clause 30.1.1.2 provides that '... the Employer is entitled to exercise any right under this contract of withholding and/or deduction...' without stating what those rights may be. In this respect JCT 98 does not differ from the pre-April 1998 version of JCT 80 and therefore preserves the employer's rights of set-off as conferred by law (see *C. & M. Pillings Ltd* v. *Kent Investments* (Chapter 1).

The requirement to state the 'grounds' on which payment is withheld does not confer an unrestricted right to deduct whatever the paying party wishes. Rather, the ground for withholding payment must amount to a set-off in law. If it does not, it should be rejected by an adjudicator (or indeed by an arbitrator or the court) on that basis alone without any investigation of the merits of the ground alleged.

The notice requirement in clause 30.1.1.4 differs from section 111 of the HGCRA only in that it uses the word 'deduction' in addition to the words 'withholding and/or'. The use of the words 'deducted' or 'deduction' do not appear to add anything to the requirement set out in section 111, save that they are consistent with clause 24 of JCT 98 (relating to the deduction of liquidated damages). It will probably be difficult to attack a notice for lack of particularity. Such attacks rarely fared well under the pre-April 1998 set-off provisions of subcontracts (see for example *Archital Luxfer Ltd* v. *A. J. Dunning* (see section 2.4)). However, in practice adjudicators may look for sufficient particularity in the notice to make it immediately obvious that the ground or grounds relied on at least arguably constitute a set-off. In those cases where the notice fails readily to demonstrate this, the more robust of adjudicators may decline to entertain the ground or grounds in question without further investigation.

Neither the JCT main contract nor subcontract standard forms expressly limit the right to set-off to claims arising out of the contract under which payment is due. In the main contract relationship this is nothing new. For delay claims the employer retains the right to withhold or deduct liquidated and ascertained damages at the specified rate. However whereas the pre-April 1998 provision involved two conditions precedent to the withholding or deduction of liquidated damages, JCT 98 imposes a three-stage process. Under the pre-April 1998 provisions of JCT 80:

- an architect's certificate of failure to complete by the completion date under clause 24.1; and
- an employer's notice on or before the date of issue of the final certificate requiring payment or deduction of liquidated damages

were conditions precedent to the deduction of liquidated damages (*J. F. Finnegan Ltd* v. *Community Housing Association Ltd* (1995)).

Under JCT 98 the entitlement to liquidated damages remains subject to the requirement for an architect's certificate under clause 24.1. However a further two stages are imposed on the employer, under clause 24.2.1:

- that the employer has informed the contractor in writing before the date of the final certificate that he may require payment of or may withhold or deduct liquidated damages; and
- that not later than five days before the final date for payment of the debt due under the final certificate the employer, in writing, shall have required (under clause 24.2.1.1) the contractor to pay or given notice (under clause 24.2.1.2) that he will deduct liquidated damages.

Failure to comply with the above criteria will disentitle the employer to liquidated damages. In practice employers are rarely likely to leave the exercise of their right to liquidated damages so late. It is more likely that they will seek to deduct liquidated damages against interim payments. Where they do so there will still have to be a valid and subsisting architect's certificate under clause 24.1 (see *A. Bell & Son (Paddington) Ltd* v. *CBF Residential Care and Housing Association* (1989)) as well as the requisite employers' notices under clause 24.2.1 and clause 24.2.1.1 or 24.2.1.2 not later than 5 days before the final date for payment of the interim certificate in question (in other words conforming with clause 30.1.1.4). In practice there appears to be no reason why the employer's notices should not be conjoined into a single notice complying in full with the requirements of clause 30.1.1.4 and clause 30.8.3. However, as a result of the stringent requirement under these latter provisions that grounds for withholding or deducting be given, it is less likely that merely dispatching a cheque for the relevant balance (as occurred in *Finnegan* v. *Community Housing*) will suffice.

For claims other than delay the employer retains the right to abate or to set off, the latter subject only to the requirement for notice in accordance with clauses 30.1.1.4 or 30.8.3. In this respect claims arising from other contracts or claims outside the contract under which payment is due may be set off provided they constitute a set-off (see Chapter 1).

It is not considered, though there may be some scope for argument, that JCT 98 permits the deduction of liquidated damages at any time before the sum in question has become due. For other kinds of loss, such as consequential damage arising from defective work, it is arguable that the employer may withhold payment in respect of future losses.

Questions may arise as to whether a notice under clause 30.1.1.4 and 30.8.3 is an 'effective notice' if it does not use the words 'proposed to be withheld...' or some other formula clearly stating what is proposed. In cases relating to the pre-April 1998 versions of the JCT and similar subcontracts, the courts adopted a relaxed approach to what constitutes notice of intention to set off (see *Archital Luxfer* v. *A. J. Dunning* and *William Cox Ltd* v. *Fairclough Building Ltd* (section 2.4), effectively allowing an implication of an intention to set off as arguably sufficient for the purposes of applications under the former Order 14 of the Rules of the Supreme Court. Adjudicators may take a similarly relaxed view of what constitutes a 'proposal' to withhold; an implication may be accepted as sufficient. What matters is substance, not form.

JCT 98 has its own adjudication procedure. Issues arise concerning the relationship between the JCT adjudication procedure and adjudication and set-off. These issues are discussed in section 3.2.1, 3.2.2, and 3.2.3.

3.3.4 Other JCT main forms

As a broad generalisation, the other JCT main forms, by the April 1998 amendments, have adopted a set-off regime similar in terms and effect to that used in JCT 98. As a result the comments on JCT 98 largely apply to those other contracts. There is some variation in the procedure for the deduction of damages for failure to complete on time, reflecting the structure of the individual contracts.

3.3.5 JCT Nominated Sub-Contract Conditions NSC/C

By virtue of Amendment 7, the NSC/C form adopts a new set-off regime (the first edition of this book concentrated heavily on the set-off provisions relating to its predecessors NSC/4 and 4A). NSC/C is similar in terms and effect to JCT 98. The subcontractor's right to payment arises under clause 4.16.1.1 in two ways:

(1) those payments due under certificates of the architect for work valued in accordance with the subcontract and included in such certificates (called Amount A) comprising the total value of the subcontract work properly executed (clause 4.17.1.1), the total value of materials and goods delivered for incorporation but not incorporated (clause 4.16.1.2) and the total value of off-site materials and goods (clause 4.17.1.3);

(2) those other 'domestic' payments due from the contractor to the subcontractor (called Amount B).

It is the architect's certificate that triggers the contractor's obligation to give a payment notice in respect of both Amount A and Amount B, and this is to be given not later than 5 days after the date of issue of the interim certificate.

Clause 4.16.1.1 then provides:

'The Contractor shall duly fulfil his obligations under clause 35.13.2 of the Main Contract Conditions in respect of Amount "A" and Amount "B".'

There is a slight oddity here because clause 35.13.2 of the main contract (whether JCT 80 or JCT 98) obliges the contractor to pay, by the final date for payment, the payments to the nominated sub-contractor directed in the interim certificates of the architect, i.e. Amount A, but not Amount B. However, it is probable that this distinction is of no real importance because the final date for payment of both Amount A and Amount B is specified as 17 days from the date of issue of an interim certificate.

Clause 4.16.1.2 of NSC/C then provides:

'Not later than five days before the final date for payment of Amount "A" and Amount "B" the Contractor may give a written notice to the Sub-Contractor which shall specify any amount proposed to be withheld and/or deducted from either Amount "A" and/or Amount "B", the ground or grounds for such withholding and/or deduction and the amount of the with-holding and/or deduction attributable to each ground.'

As a result in order to effect a set-off the contractor must give the clause 4.16.1.2 notice before the expiry of the twelfth day from the issue of an interim certificate. Despite the obligation to pay the amount certified by the architect it is considered that the contractor nonetheless retains the right to abate the sum so certified, including Amount A. This is because the basis of the architect's certification of Amount A is that the subcontract work is 'properly executed' (see clause 4.17.1.1) and that the materials have the value certified (see clause 4.17.1.2 and .2 and .3). Consequently if the work and materials are not properly executed or are defective, the contractor ought to be permitted to abate the sum certified and to dispute valuation. As the employer appears to retain the right to abate the sums certified on the basis that the valuation is disputed, it would be both wrong and unfair if the contractor were not to have a corresponding right against the subcontractor (see *C. & M. Pillings* v. *Kent Investments* (Chapter 1)).

One of the problems that troubled the draftsmen of the JCT subcontract set-off provisions – the contractor's entitlement to recover future losses – is largely unaffected by the 1998 amendments. This problem is outlined in Chapter 2 in connection with *Redpath Doorman Long* v. *Tarmac* and *Chatbrown* v. *Alfred McAlpine* where the court decided under the old Blue Form of subcontract that future losses could not be set off. Further, it was thought that the 1987 amendments to the JCT subcontracts probably shut out future loss claims from the set-off process. Clause 2.9 of NSC/C presently provides that:

> 'The Sub-Contractor shall pay or ... allow ... a sum equivalent to any loss and/or damage suffered and/or incurred by the Contractor.'

The continued use of the words '... loss and/or damage suffered or incurred...' would appear to render it unlikely that future losses can be set off for delay claims.

However, the wholesale abandonment of clauses 4.26 to 4.29 (which, presumably by a drafting oversight, is not reflected in clause 2.9) and consequent abandonment of the words '... where the Contractor has a claim for loss and/or expense and/or damage which he has suffered or incurred by reason of any breach of, or failure to observe the provisions of, the Subcontract...', as used in the former clause 4.27.1, may have altered the position at least in relation to non-delay damages claims, leaving it at least arguable that some future losses may be set off.

While the contractor has to meet a number of criteria to set-off, they are much less elaborate than the four criteria which the pre-April 1998 position imposed. For contractors' claims arising from the subcontractor's failure to complete on time, the criteria are derived from the inter-relationship between:

- clause 2.8 (notification of subcontractor's failure to complete); and
- clause 2.9 (loss and damage due to failure to complete); and
- clause 4.16.1.2 (proposal to withhold).

If the contractor wishes to set off a claim derived from delay, the criteria to be overcome are:

- a notice to the architect under clause 2.8;
- a certificate of the architect (under clause 35.15 of the main contract) certifying any failure notified under clause 2.8;
- a written notice from the contractor to the subcontractor specifying the amount to be withheld or deducted and the ground or grounds and the amount attributable to each ground.

Interestingly there is an anomaly in clause 2.9 of NSC/C in that it remains subject to clauses 4.26 to 4.29, all of which have been deleted by Amendment 7. No doubt the JCT will correct this and other anomalies in due course.

The decision in *Mellowes* v. *Snelling* (see section 2.4) remains important in relation to the NSC/C as amended in April 1998, even though the provision it concerned has been jettisoned (the case concerned clause 21.3.1.2 of NSC/4 which was initially reproduced as clause 4.16.1.2 of NSC/C). The thrust of *Mellowes* v. *Snelling* was that where the employer made deductions against the contractor, those deductions could only be passed on to the subcontractor if the employer had a right to those deductions under the main contract and if the contractor had a corresponding right under the sub-contract.

For contractors' claims for direct loss and expense under clause 4.40 the hurdles are that:

- the contractor shall have given a notice that the regular progress has been affected by an act or omission or default of the sub-contractor within a reasonable time of the effect becoming apparent;

- there must be agreement as to the amount of direct loss and/or expense which may be deducted (see *Pigott* v. *Shepherd Construction* (section 2.4));
- notice of the proposal to deduct must be given in accordance with clause 4.16.1.2.

For all other claims all that is required is a notice under clause 4.16.1.2 before withholding or deduction may be made.

Clause 4.16.1.2 gives rise to a departure from the pre-April 1998 subcontracts by permitting set-off in respect of claims not arising outside the contract itself. This is because the pre-1998 provisions contained a term to the following effect:

'The rights of the parties to the Subcontract in respect of set-off are fully set out ... and no other rights whatsoever shall be implied as terms of the Subcontract relating to set off.'

While not affecting the right to abate, this provision was found (see *Hermcrest* v. *G Percy Trentham* (section 2.4)) to restrict the contractor's right of set-off by confining it to those matters expressly allowed for in the contract (in substance claims arising under the loss and expense provisions and claims for damages for breach). Cross-claims arising from events unconnected with the contract in question could not be set off, even if notice were given. The new set-off provision arising as a result of the 1998 amendments does not maintain this restriction. As a result cross-claims arising from matters outside the contract under which payment is due may be deployed by way of set-off (see Chapter 1). In this respect the 1998 amendments to the subcontracts may come to be seen as something of a reverse by the subcontract sector of the industry. Under NSC/C, DOM/1 (and 2), NAM/SC and IN/SC it remains open to the subcontractor to assert that direct loss and expense may only be deducted by the contractor if it is 'agreed', in addition to being the subject of a notice under clauses 30.1.1.4 and 30.8.3. In this respect it is anticipated that the decision in *Pigott* v. *Shepherd* may be much cited to adjudicators and to the courts unless and until reconsidered by a higher court.

3.3.6 Other JCT subcontract forms

The JCT Works Contract 2 as amended (the amendments have not been published at the time of writing) is likely to operate a similar payment and set-off regime to NSC/C. The amendments to the other JCT subcontract forms differ from NSC/C in that payment is not triggered by a certificate. Rather it continues to be triggered by commencement of work on site, subject to any agreement in relation to off-site works. The first interim payment is due not later than one month from commencement, and subsequent interim payments are due monthly thereafter. The final date for payment is 17 days after the date when the payment became due. The employer's payment notice is to be given not later than 5 days after the due date and the notice to withhold must be given not later than 5 days before the final date for payment.

The deduction of direct loss and/or expense suffered or incurred as a result of failure to complete on time is dependent on the contractor giving notice to that effect. Like NSC/C the contractor's entitlement to deduct 'agreed' direct loss and/or expense caused by a material effect on regular progress resulting from act, omission or default of the subcontractor is dependent on a notice to withhold being given, It also begs the question whether direct loss and expense which is not agreed can be deducted at all (see *Pigott* v. *Shepherd* (section 2.4)).

3.4   Other standard forms

3.4.1   The ICE Conditions 6th Edition

As with other standard main forms the ICE Conditions 6th Edition, prior to the advent of the HGCRA, did not place any restriction on the employer's rights of set-off as conferred by law. Further, clause 66 of the Conditions created a dispute procedure under which disputes had first to be referred to the engineer for a decision, whereupon if the engineer failed to give a decision or the decision was itself disputed then the dispute could be referred to arbitration. This procedure did not fetter the employer's right to abate or set-off against a disputed certificate (see *Enco Civil Engineering Ltd* v. *Zeus International Development Ltd*).

The Institution of Civil Engineers responded to the HGCRA in 1998 by amending the ICE Conditions 6th edition. Interim payment remains dependent on a certificate of the engineer. Certificates are, pursuant to clause 60(2), to be issued 25 days after the contractor has submitted a monthly statement of the value of the work executed.

The certified payment becomes due on the date of the certificate, and the final date for payment is 28 days after the date of delivery of the contractor's monthly statement. By the redrafted clause 60(9), the certificate constitutes the payment notice for the purposes of section 110(2) of the HGCRA.

Clause 60(10) has been redrafted as follows:

> '(10) Where a payment under Clause 60(2) or (4) is to differ from that certified or the Employer is to withhold payment after the final date for payment of a sum due under the Contract the Employer shall notify the Contractor in writing not less than one day before the final date for payment specifying the amount proposed to be withheld and the ground for withholding payment or if there is more than one ground each ground and the amount attributable to it.'

Once again it is submitted that:

- the employer may abate without giving notice; and
- the ground for withholding payment as specified in the employer's notice must amount in law to a set-off.

Clause 66 has been substantially amended. It now provides by subclause 66(2) for matters of 'dissatisfaction', rather than disputes, to be referred to the engineer. By clause 66(3) it is provided that '...no matter shall constitute nor be said to give rise to a dispute unless and until...' the time for the engineer to give his decision under clause 66(2) has expired without him having done so, or the decision is unacceptable or not complied with, or an adjudicator shall have given a decision which either party is not giving effect to. A notice of dispute must be served on the other party and the engineer before a dispute will be said to exist. Only then may the matter be referred either to adjudication, conciliation or arbitration.

There is some debate as to whether this provision might contravene the requirements of section 108(2)(a) of the HGCRA on the basis that the definition of dispute purports to inhibit the right to refer disputes to adjudication 'at any time'. It might be asked, however, how there can be grounds to withhold payment if there is no dispute until the above process has been exhausted! The answer seems to be that, if the employer exercises his right to withhold payment under clause 60(10), there is 'no dispute' about his right to do so until the provisions of clause 66(3) have been satisfied.

3.4.2 Other standard conditions of contract

A variety of other standard forms of main and subcontract conditions have been amended to take account of the HGCRA. The thrust of most amendments is that the right to abate (subject to the effect of the adjudicator's decisions) remains intact and set-off may be effected subject to the giving of an effective notice that complies with the requirements of the HGCRA as regards timing and content. Further, the New Engineering Contract and Subcontract follow the approach adopted by the amendments to ICE Conditions in relation to the deferment of the existence of disputes until after certain procedures have been undertaken.

3.4.3 Consultants

As mentioned earlier, through section 104(2) of the HGCRA, contracts for architectural, design or surveying work and the provision of advice on building engineering, decoration and land-scaping in relation to construction operations, all constitute construction contracts. As a result, contracts entered into by architects, engineers, surveyors (whether quantity, building or valuation surveyors) may constitute construction contracts. Various bodies have issued amendments to their standard terms of engagement as a result, e.g. the Association of Consulting Engineers, the Institution of Civil Engineers and the Royal Institute of British Architects. None of these conditions restrict the exercise of the common law right to abate (see, however, *Hutchinson* v. *Harris*), and payment may be withheld provided the ground relied on constitutes a set-off in law and is the subject of an effective notice.

CHAPTER FOUR
THE POSITION UNDER STANDARD TERMS OF BUSINESS

4.1 Introduction

Written standard terms of business conditions frequently contain provisions which purport to exclude the ordinary rights of set-off otherwise available to one of the parties to the contract. Some old authorities suggested that a right of set-off cannot be waived. However, in *Hong Kong & Shanghai Banking Corporation* v. *Kloeckner A.G.* (1989), the High Court (in connection with a case concerned with letters of credit) refused to follow those old cases and preferred a more recent decision, *Halesowen Presswork and Assemblies Limited* v. *Westminster Bank Limited* (1971), that the right can be excluded by agreement, express or implied. Latterly, in *Coca Cola Financial Corporation* v. *Finsat International Limited and Others* (1996) the Court of Appeal rejected the proposition that section 49(2) of the Supreme Court Act 1981 or public policy prevented a party from excluding the right of set-off.

4.2 The Unfair Contract Terms Act 1977

The type of provision described above may be void if it fails the test of reasonableness within the meaning of section 13(1)(b) of the Unfair Contract Terms Act 1977 which provides:

'(1) To the extent that this Part of this Act prevents the exclusion or restriction of any liability it also prevents:

 . . .

(b) excluding or restricting any right or remedy in respect of the liability, or subjecting a person to any prejudice in consequence of his pursuing any such right or remedy;'

Stewart Gill Ltd v. Horatio Myer & Co Limited (1992) 9-CLD-02-09
In this case the Court of Appeal had to consider the effect of section
13 (1) (b).

The facts

Myer entered into a contract with Gill for the supply and installation of a conveyor system for £266,400 to be paid in stages, the final
10% being payable by 5% on completion and 5% 30 days thereafter.
The contract incorporated Gill's conditions of sale, clause 12.4 of
which provided that:

> 'the customer shall not be entitled to withhold payment of any
> amount due to the company under the contract by reason of any
> payment credit set-off counterclaim allegation of incorrect or
> defective goods or for any other reason whatsoever which the
> customer may allege excuses him from performing his obliga-
> tions hereunder.'

On completion of the installation Myer withheld the final 10%
alleging breaches of contract by Gill giving rise to cross-claims
which could be set off against the amount claimed. Gill commenced
proceedings and applied for summary judgment for the amount
claimed, relying on clause 12.4 of the contract. The judge refused
Gill's application for summary judgment, holding that clause 12.4
was subject to section 13 of the Unfair Contract Terms Act 1977 and
could not be relied on because it was unreasonable. Stewart Gill
appealed.

Held

The Court of Appeal upheld the judge's decision on the basis that a
term in a contract preventing a payment or credit being set off
against the price claimed was prima facie unreasonable, therefore
the burden of satisfying the court that the term was 'reasonable' and
not rendered ineffective by section 13 of the 1977 Act rested on the
party relying on the term. Furthermore, the question of whether a
contract term satisfied the requirements of reasonableness in section
13 of the 1977 Act had to be determined by considering the term as a
whole and not merely that part of it relied on to defeat the set-off. It
followed therefore that the whole of clause 12.4 was unenforceable.
Consequently, Gill was not entitled to summary judgment.

The *Stewart Gill* case was concerned with a contract where one party dealt with the other on the other's written standard terms of business. The effect of the *Stewart Gill* case, i.e. the application of the Unfair Contract Terms Act to provisions restricting rights of set-off, may be far more wide ranging than might at first appear; it may apply even where the parties' contract uses a standard form contract because of the broad interpretation given to written 'standard terms of business' by Judge Stannard QC in *Chester Grosvenor Hotel Company Ltd* v. *Alfred McAlpine Management Limited and Others* (1991).

Chester Grosvenor Hotel Company Ltd v. Alfred McAlpine Management Ltd and Others (1991) 56 BLR 115

The facts

Chester engaged McAlpine as management contractors under two separate contracts for work to the hotel. The form of management contract was devised by McAlpine and contained provisions restricting their liability, which they sought to rely on when disputes arose under the contract. McAlpine's management contract could not be directly compared with the JCT management contract.

Held

In considering whether the management contract fell within the meaning of 'written standard terms of business', Judge Stannard cited the considerations that should apply in the following terms:

> 'I accept that where a party invariably contracts in the same written terms without material variation, those terms will become its "standard form contract" or "written standard terms of business". However, it does not follow that because terms are not employed invariably, or without material variation, they cannot be standard terms. If this were not so the statute would be emasculated, since it could be excluded by showing that, although the same terms had been employed without modification on a multitude of occasions, and were employed on the occasion in question, previously on one or more isolated occasions they had been modified or not employed at all. In my

judgment the question is one of fact and degree. What are alleged to be standard terms may be used so infrequently in comparison with other terms that they cannot realistically be regarded as standard, or on any particular occasion may be so added to or mutilated that they must be regarded as having lost their essential identity. What is required for terms to be standard is that they should be regarded by the party which advances them as its standard terms and that it should habitually contract in those terms. If it contracts also in other terms, it must be determined in any given case, and as a matter of fact, whether this had occurred so frequently that the terms in question cannot be regarded as standard, and if on any occasion a party has substantially modified its prepared terms, it is a question of fact whether those terms have been so altered that they must be regarded as not having been employed on that occasion.'

This broad interpretation may have implications not only for the 'small print' on the reverse of conditions of sale and purchase but will also affect 'in house' or 'tailor made' standard forms and may even extend to the use of industry-recognised standard forms where one party habitually relies on a particular form.

4.3 Standard terms and the HGCRA

Standard terms of business, where they are incorporated in or give rise to a 'construction contract', are subject to the provisions of the HGCRA. Where the standard terms fail to make provision for the matters provided for in the HGCRA, the relevant parts of the Scheme will be implied.

It should be noted that the HGCRA does not prevent the exclusion of the right of set-off; rather, by section 111 it provides that the exercise of the right to set-off is subject to the giving of an effective notice. However, where standard terms of business go further than the implied terms under the HGCRA, for example by excluding the paying party's rights of set-off altogether or by subjecting them to exceptionally onerous notice requirements, that may give rise to the relevant term being declared unreasonable and therefore unenforceable under the Unfair Contract Terms Act. Adjudicators under the HGCRA could therefore find themselves asked to embark on the difficult task of determining whether the term in question is unfair. Where a construction contract comprises standard terms of business which exclude altogether or onerously restrict the ordinary rights of set-off, the paying party ought, without prejudice to the exclusion clause, to give notice of any set-off to be effected. If they do not, in the event that the paying party succeeds in having the exclusionary term declared unenforceable, the set-off will none-theless remain invalid due to the absence of an effective notice.

4.4 Standard terms and consumers

A further complication arises where a contract is entered into by a person who is a 'consumer' for the purposes of the Unfair Terms in Consumer Contracts Regulations 1994 (SI 1994/3159). A 'consumer' is defined in the Regulations as 'a natural person who, in making a contract to which these regulations apply, is acting for purposes which are outside his business...'. 'Business' includes trade or profession (see regulation 2(1)).

The consumer could enter into a contract which contains terms conforming with the HGCRA, although conformity is not compulsory. This can occur where the contract in question is excluded from the definition of 'construction contract' (see HGCRA section 104 (4)), or where the work does not come within or is excluded from the definition of 'construction operations' (see HGCRA section 105), or where the work is being carried out on a dwelling for a person (the consumer) who is a residential occupier (see HGCRA section 108). In these circumstances there is a limited opportunity available for the consumer to avoid the effect of the HGCRA restrictions on the ordinary rights of set-off available at law.

By way of illustration, the 'consumer' may enter into a contract with a builder for work in connection with a 'dwelling' (see HGCRA section 106 (2)). At the suggestion of the builder the JCT Agreement for Minor Building Works MW 98 (the successor to the 1980 Minor Works Agreement) is used. The MW 98 agreement contains provisions restricting the right of set-off so as to conform with the HGCRA. Ordinarily a residential occupier (likewise an employer where the contract is not a construction contract or does not involve construction operations) could refuse to accept any terms conforming with the HGCRA. In this illustration, however, the consumer is ignorant of his right to refuse the HGCRA conforming terms. In addition, the builder fails to complete on time and is therefore liable to pay compensation for loss by way of liquidated damages, but the consumer is again ignorant of the requirement to give an effective notice of deduction in accordance with the provisions of MW 98. Assume then that the builder commences proceedings to recover the price and seeks to avoid his liability for liquidated damages by relying on the absence of an effective notice. In these circumstances the consumer will, prima facie, be able to rely on the regulations which by schedule 3 indicate that a term 'inappropriately excluding or limiting the legal rights of the consumer . . . including the option of offsetting a debt . . .' may be regarded as unfair.

The consumer may rely on the regulations because they provide that any terms of a contract:

- which '. . . have not been individually negotiated' (see regulation 3 (1)), *or*
- where the consumer '. . . has not been able to influence the substance of the term' (see regulation 3 (3)), *or*
- where '. . . an overall assessment of the contract indicates that it is a pre-formulated standard form' (see regulation 3 (4)), *and*
- which are unfair, i.e. '. . . contrary to the requirement of good faith causes a significant imbalance . . . to the detriment of the consumer' (see regulation 4 (1)

are not 'binding on the consumer') (see regulation 5 (1)).

Schedule 2 of the regulations sets out criteria to be considered in the assessment of good faith and schedule 3 sets out an indicative and illustrative list of terms which may be regarded as unfair.

Consumers who enter into contracts which are not subject to the mandatory requirements of the HGCRA may not therefore be bound by those terms of the contract which restrict what would otherwise be their rights of set-off at law.

Consumers who enter into contracts which are subject to the mandatory provisions of the HGCRA are unlikely to be able to obtain the protection of the regulations. This is because it is provided in schedule 1 (e) (ii) of the regulations that they do not, inter alia, apply to '. . . any term incorporated in order to comply with or which reflects . . . statutory or regulatory provisions of the United Kingdom . . .'. In this respect schedule 1 (e) (ii) prima facie appears to conform with Article 1 (2) of the EU Directive (Council Directive 93/13/EEC–O.J. number L95, 21.4.93, p.93) which the regulations are intended to implement. As a result a challenge to the validity of schedule 1 (e) (ii) on the basis that it results in a failure by the UK government to implement the directive, is unlikely. There is, however, some conflict between Article 1 (2) and the twelfth recital of the regulations which provides:

'Whereas the statutory or regulatory provisions of the Member States which directly or indirectly determine the terms of consumer contracts are presumed not to contain unfair terms...'.

The recital appears to envisage a rebuttable presumption that statutory or regulatory provisions are not unfair. That presumption has not been translated in Article 1 (2). Whether this opens the way to a challenge to schedule 1 (e) (ii) of the regulations is an open question.

CHAPTER FIVE
THE POSITION UNDER PERFORMANCE BONDS

5.1 Introduction

Performance bonds are a common feature of the construction process. Almost invariably they consist of a tripartite agreement under which the bondsman binds himself for payment of a given sum if a contract is not performed or if there is default in the payment of damages.

Bonds have tended to fall into two distinct categories: those that are payable 'on demand' or those that are payable 'on proof of conditions'. The category into which a particular bond falls into will depend on its wording; however, a form of bond evolved which used wording that had become traditional, albeit archaic. The traditional wording had been subject to criticism in *Trade Indemnity Company Ltd* v. *Workington Harbour & Dock Board* (1937) where Lord Atkin remarked:

> 'it is difficult to understand why businessmen persist in entering upon considerable obligations in old fashioned forms of contract which do not adequately express the true transactions.'

The traditional bond had long been assumed to be payable on proof of conditions. However, the industry was thrown into confusion for a time as a result of the decision of the Court of Appeal in *Trafalgar House Construction (Regions) Ltd* v. *General Surety & Guarantee Company Ltd* (1994) which decided that the purpose of the traditional bond was the creation of an obligation on the part of the surety to provide funds once a demand was made, the payment being the amount of the damages asserted in good faith by the party calling the bond.

5.2 The traditional wording

The wording of the bond with which the Trafalgar House was concerned was in the following terms:

'BOND
BY THIS BOND WE ... ("the Sub-Contractor") ... and GENERAL SURETY & GUARANTEE CO LIMITED (hereinafter called "the Surety") are held and firmly bound unto ... "the Main Contractor" ... in the sum of £101,285.00 ... for the payment of which sum the Sub-Contractor and the Surety bind themselves their successors and assigns jointly and severally by these presents
NOW THE CONDITION of the above written Bond is such that if the Sub-Contractor shall duly perform and observe all the terms provisions conditions and stipulations of the said Sub-Contract on the Sub-Contractor's part to be performed and observed according to the true purport intent and meaning thereof *or if on default by the Sub-Contractor the Surety shall satisfy and discharge the damages sustained by the Main Contractor hereby up to the amount of the above-written Bond* then this obligation shall be null and void but otherwise shall be and remain in full force and effect...'

Trafalgar House Construction (Regions) Ltd v. General Surety and Guarantee Co Ltd (1995) 73 BLR 32, HL

The facts

On 8 September 1989 Trafalgar as main contractor contracted for the construction of a new leisure complex for £9m. Trafalgar engaged a subcontractor for groundworks. Trafalgar required the subcontractor to provide a bond for an amount equal to 10% of the subcontract value. An acceptable bond, worded as set out above, was executed by the subcontractor and General Surety. Before completion of their work the subcontractor went into receivership and were unable to continue. Trafalgar therefore took over completion of the work and then advanced a claim under the bond and obtained summary judgment on 9 February 1993 on the grounds that there was no arguable defence to the claim.

Held

In the Court of Appeal, Saville LJ observed that little had changed after 60 years following the criticism in the *Trade Indemnity* case and that 'the court is faced with the problem of trying to ascertain from the archaic and unsatisfactory language the parties have chosen to use what bargain they in fact intended to make.'

It was concluded that the bond imposed on the surety an independent obligation to pay the damages sustained by Trafalgar (up to the amount of the bond) from a failure of the subcontractor to carry out the subcontract and that the obligation to pay arises when called upon to do so by Trafalgar. In this respect, in the Court of Appeal, the surety argued that what had to be paid was the 'net' amount due from the subcontractor to Trafalgar after taking into account all debits and credits (including the value of any set-offs, cross-claims and the like).

The Court of Appeal disagreed. In effect the Court of Appeal, contrary to popular wisdom, regarded the traditional wording as creating an 'on demand' obligation. General Surety appealed.

The decision of the Court of Appeal in the *Trafalgar House* case was reversed by the House of Lords on 29 June 1995. The House of Lords held that:

- on its true construction the bond amounted to a guarantee so that General Surety were entitled to raise all questions of sums due and cross-claims which would have been available to the subcontractor in an action against them for damages; and
- as a matter of construction, the words 'damages sustained by the main contractor thereby' as used in the bond did not define the appellant's obligations solely by reference to the additional expenditure incurred by Trafalgar without reference to any sums that could normally be set off against them in an action for damages against the subcontractor; and
- clear and unambiguous language was necessary to displace the normal legal incidents of suretyship so that proof of damage and not mere assertion thereof was required before liability arose; and
- the bondsman was entitled to unconditional leave to defend on the grounds that there was sufficient evidence to raise a triable issue as to the heads of claim; and
- in any event some other reason for trial existed within the meaning of Order 14 rule 3 of the Rules of the Supreme Court

because there would have to be a trial of the subcontractor's counterclaims which were capable of being set off.

More recently, Rix J, sitting in the Commercial Court, in *BOC Group v. Centeon LLC* (1998) (see Chapter 1) decided (without *Trafalgar House* being cited to him) that a guarantor could avail himself of any right of set-off available to the primary debtor unless the factual background was, or the express wording of the guarantee had the effect of, abrogating the general rule. He also decided that the guarantor could rely on any cross-claim of their own against the condition provided the express terms of the guarantee did not preclude the exercise of this right.

5.3   Modern wording

The decision of the House of Lords ultimately justified the inertia that had maintained the use of the traditional bond wording in spite of the criticisms. It is clear, however, that the adoption of wording other than that used in the *Trafalgar House* case may itself be subject to scrutiny and may raise questions as to the status of obligation, i.e. is it a guarantee and if so are the rights of set-off normally attaching to a guarantee in any way inhibited? The Joint Contracts Tribunal in drafting its new Advance Payment Bond and Bond in respect of payment for off-site materials and goods (see, for example, JCT 98) has included a specific clause aimed at inhibiting the exercise by the bondsman of the right of set-off, as set out in clause 4 and 6 respectively of the bonds:

'Payments due under this Bond shall be made notwithstanding any dispute between the Employer and the Contractor and whether or not the Employer and the Contractor are or might be under any liability one to the other...'

5.4   Bonds and the HGCRA

As mentioned in section 3.1.2 performance bonds are excluded from the scope of the HGCRA.

CHAPTER SIX
SUMMARY JUDGMENT, INTERIM PAYMENTS AND SET-OFF

6.1 Introduction

Procedures exist in the High Court whereby a party may make application for either summary judgment or an interim payment to be made pending the trial of the action. The prevalence of arbitration clauses in building and engineering contracts combined with the effect of Section 9 of the Arbitration Act 1996 appeared likely, at one time, to render the court procedures relevant only in rare cases. This view was reinforced by the decision of the Court of Appeal in *Halki Shipping Corporation* v. *Sopex Oils Ltd* (1998). However, since then two important developments have occurred. Firstly, the JCT by its 1998 amendments to its Forms of Contract has introduced an optional arbitration clause and has excluded adjudication enforcement from the scope of its standard arbitration clause. Secondly, *Halki* has been distinguished by the first instance decision in *Macob* v. *Morrison* (1999). The pendulum is therefore swinging back in favour of litigation rendering an understanding of the relationship between the courts' powers, arbitration and adjudication all the more important.

The High Court procedures are now contained in Part 24 and Part 25 respectively of the Civil Procedure Rules, which come into effect on 26 April 1999 replacing the former Orders 14 and 29. Ordinarily, parties in High Court proceedings will have to proceed to trial prior to obtaining payment of monies due to them. This can often be a lengthy and costly procedure. However, the summary judgment and interim payment procedures confer on a party the right to obtain payment earlier where the merits of a party's case give rise to no real prospect of the defence succeeding, in the case of Part 24, or if the court can be satisfied that the claimant will after a trial obtain judgment for a substantial sum, in the case of Part 25.

All the cases dealt with in section 2.4 involved applications for summary judgment under Orders 14 or 14A.

6.2 *Summary judgment*

The summary judgment procedure is governed by Part 24 CPR. The requirements of Part 24 are that:

(1) the defendant must have acknowledged service of the claim form or filed a defence unless the court gives permission or a practice direction provides otherwise (see Rule 24.4);
(2) the defendant has no real prospect of succesfully defending the claim or a particular issue (see Rule 24.2); and
(3) there is no other reason why the case or issue should be disposed of at a trial (see Rule 24.2).

An application for summary judgment under Part 24 must be supported by written evidence showing that the claimant is entitled to judgment. If the claim is unanswered by the defendant or if the defendant fails to show any other reason for a trial the court will give judgment. Where it appears possible that a claim or defence may succeed but improbably that it will do so the court may make a conditional order.

The application is normally heard in chambers before a Registrar or Master of the High Court. However, if the matter is commenced in the Technology and Construction Court (formerly the Official Referee's court), as most building cases are, the hearing will be before a judge. Judgment will be given if the claimant shows an unanswered claim and if no reason for a trial exists. If a conditional order is made it may require money to be paid into court or that specified steps be taken.

The case of *Enco Civil Engineering Limited* v. *Zeus International Development Limited* (1991) illustrates the difficulties a plaintiff may encounter when the evidence in support of the Part 24 application is flawed and remains relevant. In this case, in relation to a certificate of the engineer (under the ICE Conditions 5th Edition) certifying payments as due, Judge Esyr Lewis QC declined to enter judgment under Order 14 as it then was, on the technical basis that there was no evidence before the court to show when the monthly statement (required under Clause 60(1)) which founded the certificate was delivered to the engineer, or to show when the 28 day period from that time expired, so that the plaintiffs had failed to establish a necessary ingredient in their claim for final judgment.

The case of *Pigott* v. *Shepherd* (1993) is illustrative of an Order 14 defeat snatched from the jaws of victory. The plaintiff succeeded in establishing, on an application under Order 14A, that on the proper construction of the contract – a DOM/1 standard form incorporating a number of ad hoc amendments – the sum of £160,086.33 was due. However, by a very late amendment the defendant raised a claim to have the agreement rectified to delete a provision limiting damages for delay to £40,000. The judge regarded the claim to rectification as arguable and as a result the plaintiff was unable to obtain judgment or an interim payment.

In situations where there may be some defence but such a defence is still open to doubt, the defendant may be allowed to defend on terms that either the whole or part of the claim is paid into court to abide the event. This order (rare in construction disputes), when made, will not allow the monies to be paid to the plaintiff but can often lead to a settlement of the dispute as any cash flow advantage to the defendant will have been removed or reduced.

While it is often open to some considerable argument whether there is a bona fide defence to many claims, and while there can be no certainty of success, many claimants are tempted to take steps to apply for summary judgment on the ground that, if successful in whole or in part, this will avoid a long and costly procedure of having to proceed to a full trial of the merits. It can change fundamentally the course of a dispute between the parties.

In the Court of Appeal decision in *United Overseas Ltd v. Peter Robinson Ltd* (1991), Bingham LJ analysed the authorities and found that there are four categories of order which should be made when a defendant raises a set-off or a counterclaim:

(1) where the defendant can show an arguable set-off, whether equitable or otherwise, he is entitled to leave to defend to the extent of the set-off and the court has no discretion;

(2) where the defendant sets up a bona fide counterclaim arising out of the same subject matter as the action and connected with the grounds of defence, the order should not be for judgment on the claim, subject to a stay pending trial of the counterclaim, but should be for unconditional leave to defend, even if the defendant admits the whole or part of the claim;

(3) where there is no defence to the claim but a plausible counterclaim of not less than the claim is set up, judgment should be for the plaintiff on the claim with costs, stayed until trial of the counterclaim;

(4) where the counterclaim arises out of a separate and distinct transaction or is wholly foreign to the claim, judgment should be for the plaintiff with costs without a stay.

While the terminology and procedural options have changed under Part 24 of the CPR the substance of the Robinson approach will continue to apply. It should also be borne in mind that these *Robinson* principles may not generally be applicable where the application for summary judgment seeks to enforce the decision of an adjudicator. A more detailed commentary on this aspect is set out in sections 3.2.2 and section 3.2.3.

The procedure for applications for summary judgment are set out in Part 24 of the CPR and the relevant Practice Direction (see Appendix 8).

An application for summary judgment must be supported by written evidence deposing to all the material facts relating to the questions to be determined by the court. The defendant may file written evidence in answer. If he does so, disputing the facts, that will generally constitute a reason for ordering a trial. Even if the application relates to an issue of law or construction, if the evidence challenges or contradicts the facts as deposed to by the claimant, there is unlikely to be any scope for the court to give judgment or make a conditional order since there is no room for dispute as to the necessary material facts (see for example *Imperial Square Developments (Hoxton) Limited* v. *Aegon Insurance Co (UK) Limited* (1998).

The Part 24 procedure nonetheless enables the court to deal with a question of law or construction of a document at any stage of the proceedings where it appears that such a question is suitable for determination without a full trial of the action and that such determination will finally determine the entire cause or matter or claim or issue. Examples of the use of the Order 14A procedure (as it then was) may be found in the cases of *Colbart Ltd* v. *H. Kumar* (1992) (concerned with the interpretation of JCT Intermediate Form IFC 84 clauses 1.1, 1.10, 4.7, 6.3c and 6.8) and *Pigott Foundations Ltd* v. *Shepherd Construction Ltd* (1993) (concerned with the interpretation of clause 11.1, 11.8 and 13.4 of DOM/1). These cases could equally be disposed of under the new procedure.

6.2.1 Summary judgment and arbitration agreements

Arbitration has long been recognised as the accepted means of resolving disputes in the construction industry and clauses exist in the forms of contract published by the Joint Contracts Tribunal and other bodies for the referral of disputes to an arbitrator.

Accordingly, it became the common practice employed by

defendants when served with a writ to make an application to the court to stay the High Court proceedings on the grounds that there was an arbitration clause in the contract whereby the parties had agreed that any dispute between the parties should be referred to an arbitrator. Prior to February 1997 such an application was made under the Arbitration Act 1950, section 4. When dealing with arguments as to whether a right of set-off exists, the courts had no difficulty, at one time, in justifying their right to hear Order 14 applications since the courts had held that a prerequisite for the operation of the arbitration clause must be the existence of a genuine dispute or difference between the parties.

The whole basis of an application for summary judgment under Part 24 CPR remains in effect that there is no triable issue giving rise to a dispute or difference and, hence, no real prospect of the defence succeeding. In the event that a Part 24 application is unsuccessful, by implication a dispute or difference exists, and thereafter the court will give due weight to the arguments for the stay of the proceedings on the basis that the subject matter should be referred to arbitration.

However, prior to the advent of the Arbitration Act 1996 there were conflicting authorities as to whether it was correct to inquire into the genuineness of the dispute. (See *Hayter* v. *Nelson & Others* (1990), *Mayer Newman & Co. Ltd* v. *A1 Ferro Commodities Corporation SA* (The John C. Helmsing) (CA) (1990), *R. M. Douglas Construction Ltd* v. *Bass Leisure Ltd* (1990) and *John Mowlem & Co Plc v. Carlton Gate Development Co. Ltd* (1990)).

The Arbitration Act 1996 applies to all arbitrations commenced after January 1997. It obliges the court, pursuant to Section 9(4), provided the application is made before the applicant has taken any step to answer the substantive claim, to grant a stay of any matter which under the arbitration agreement is to be referred to arbitration. The court will not grant a stay where the arbitration agreement is null and void, inoperative or otherwise incapable of being performed, and it is not obliged to grant a stay when the agreement is subject to the provisions of sections 89 to 91 of the Arbitration Act 1996.

Shortly after the Arbitration Act 1996 came into force, the court was invited to consider whether it should enquire into the genuineness of a dispute. This occurred in *Halki Shipping Corporation* v. *Sopex Oils Limited* (1998) where the Court of Appeal, by a majority, upheld the decision of Clarke J. When dealing with cross applications under Order 14 of the Rules of the Supreme Court and section 9 of the Arbitration Act 1996, Clarke J concluded that section 9(4) had the effect, except in very unusual circumstances, that:

- disputes within the arbitration clause were to be referred to arbitration;
- a dispute arising from or in connection with the contract included any claim by one party which the other party refused to admit or did not pay; and
- even claims to which there was obviously no answer in fact or in law were no longer justiciable by legal process but were bound to be referred to arbitration.

In the Court of Appeal it was concluded that there was a dispute and that the arbitration agreeement in question did not limit the disputes capable of being referred to arbitration solely to those that were not capable of being resolved by summary judgment procedures. Henry LJ put it this way:

'... I would not be immediately impressed by a submission that I should construe "dispute" with so artificial a narrowness as to be restricted to such disputes (as to liability or quantum) as are found by the court to merit the grant of leave to defend – after a contested hearing for summary judgment under Order 14, which often takes hours and sometimes takes days ... to put it another way, when the parties have chosen arbitration for their dispute resolution, I would not (if unconstrained by statute or authority) interpret their choice as being restricted to referring only to those disputes that cannot be resolved by the court's summary judgment procedures.'

Swinton-Thomas LJ put it as follows:

'There is a dispute once money is claimed unless and until the Defendants admit that the sum is due and payable, and if a party has refused to pay a sum which is claimed or has denied that it is owing then in the ordinary use of the English language there is a dispute between the parties, and the Court no longer has to consider whether there is <u>in fact</u> any dispute between the parties but only whether there is a dispute within the arbitration clause of the agreement.'

Despite a powerful dissenting judgment by Hirst LJ, the decision in *Halki* represents a formidable obstacle to the use of summary judgment where it is alleged that a dispute exists within the meaning of an extant arbitration agreement. Matters are not helped by the subsequent decision in *Ahmad Al-Naimi (trading as Buildmaster Construction Services)* v. *Islamic Press Agency Incorporated* (1998) where it was decided that a question of construction of an arbitration agreement should not be decided by the court. However, *Macob* v. *Morrison* distinguishes *Halki* where the court is concerned with an adjudicator's decision.

The existence of a binding arbitration agreement may, however, have the effect of putting an obstacle in the way of a swift decision. Where it does, this begs the question as to whether a stay should be granted if under the arbitration agreement the dispute cannot immediately be referred to arbitration. This issue arose in the case of *Enco Civil Engineering Limited* v. *Zeus International Development Limited* (1991) in relation to a cross application under section 4 of the Arbitration Act 1950 for a stay of the action to arbitration. It was decided that although disputes could not at the time be referred to arbitration because of the two-stage procedure prescribed by clause 66 of the ICE Conditions 5th Edition, this did not prevent the court from ordering a stay of the action.

The *Enco* decision was approved by the Court of Appeal in *The Channel Tunnel Group Limited & Another* v. *Balfour Beatty Construction Limited and Others* (1992) and, implicitly, by the House of Lords. This principle has been adopted in section 9(2) of the Arbitration Act 1996.

The Part 24 procedure has the advantage of providing a claimant with a speedy means of redress against a defendant who is alleged to have no right of set-off against the claimant's claim. Where the contract provides for the resolution of all disputes by arbitration, claimants may be prejudiced by the denial of such a speedy means of redress unless the arbitrator is empowered under section 39 of the Arbitration Act 1996 to grant provisional orders, or can otherwise deal with the matter under section 47 of that Act.

Both these provisions present difficulties. The power under section 39 to grant provisional relief may only be exercised where the parties agree to confer this power on the arbitrator, which may prove a significant restriction on its use. (Certain arbitration rules, such as the JCT 1998 Edition of the Construction Industry Model Arbitration Rules (CIMAR) expressly empower the arbitrator to grant provisional relief under section 39). Where section 39 does, however, apply, it would appear that arbitrators may grant interim financial relief akin to the court's power to make interim payment orders under Order 29 of the Rules of the Supreme Court (now Part 25 CPR). If this is so, the arbitrator may be constrained in the same manner as the court (see section 6.3.1), in which case section 39 even when it applies may be of limited value to claimants.

The power under section 47 enables the arbitrator to make final awards on discrete issues. Unlike section 39 such awards are not provisional and so may not be reviewed by the arbitrator at a later stage in the proceedings. The power may be used, however, to determine rights of set-off, e.g. where a respondent deploys an alleged set-off or abatement, but the grounds relied upon do not in law give rise to abatement or set-off. If not it may be appropriate to make an award in favour of the claimant in respect of the whole or some part of the claim.

6.2.2 Summary judgment and adjudication

Even in relation to those contracts which require all disputes to be referred to arbitration, the Part 24 procedure will be available as one of a range of procedures for securing compliance with the decision of an adjudicator for the payment of money (see *Macob* v. *Morrison*). The JCT standard forms of main and subcontract, as amended in April 1998, expressly exclude the enforcement of adjudicators' decisions from the operation of the (now optional) arbitration agreement. The Part 24 procedure entitles the defendant to defend where a defence with real prospects of success is established. As explained in Chapter 3, claimants are bound to argue that the word 'binding' as used in section 108 of the HGCRA – and the words 'binding and shall comply with' as used in Rule 23 of the Scheme – do not admit any defence to an adjudicator's decision.

6.3 *Interim payments*

The former Order 29 Part II gave rise to a discretion on the part of the court if it thought fit, and without prejudice to any contentions of the parties, to order a defendant to make an interim payment of such amount as it considered just after taking into account any set-off, cross-claim or counterclaim on which the defendant may be entitled to rely. One of the grounds for making such an order was that the court believed the plaintiff would obtain judgment against the defendant for a substantial sum of money if the action were to proceed to a trial. Thus the court, on an interlocutory basis, was making a reasoned assessment of the outcome of any trial. Part 25 of the CPR adopts the same approach (see Appendix 9).

An application for an interim payment may now be made at any time following the filing of an acknowledgment of service on the part of a defendant. However, a court may not view an unexplained delay in the making of an application with any great favour. Application is made to a registrar, master or judge if the action is commenced in the Technology and Construction Court. Often, the application for an interim payment will be combined with an application seeking summary judgment under Part 24. Any application for an interim payment must be supported by written evidence setting out the grounds for application and the facts relied upon.

It is important to note that an interim payment, even in the unusual circumstances where the sum ordered to be paid equals the full amount of the plaintiff's claim, is no substitute for the trial process itself. The dispute may continue and a trial between the parties may be heard. The interim payment procedure is not an unconditional order of the court and may subsequently be varied. It is a payment on account in the expectation that the claimant will succeed at the trial for a substantial amount. This may be contrasted with the procedure for summary judgment under Part 24 where the order, once made, is final unless it is the subject of appeal.

Consideration of cases under Order 29 RSC shows how applications for interim payment, where an allegedly wrongful set-off occurs, will often be of no avail.

Those familiar with construction industry disputes will recognise the following scenario. A subcontractor's work departs from the specification but still functions. However, it may or may not lead to increased future maintenance and it increases the risk of problems which might not otherwise have resulted. Because of this the main contractor will not treat the subcontractor's work as having been practically completed. However, the employer, who is aware of the departure, forms the view that it is not necessary for the work to be put right. This would cause significant dislocation to the employer and involve very great expense to the subcontractor and/or main contractor. The employer, therefore, issues a certificate of practical completion of the works but does not accept the departure from the specification as a variation. The employer then pays the main contractor in respect of the subcontract work. The employer expressly reserves his rights against the main contractor should there be future maintenance problems or should the work result in a functional failure. The main contractor is a solid well-funded organisation and the employer is happy to rely on this fact.

However, the main contractor does not regard the subcontractor as being similarly solid and well-funded.

If the main contractor some years hence is required to put right or to pay for a functional failure in the subcontractor's work, there is a risk that the subcontractor will no longer be around, and this would leave the main contractor bearing the cost of rectification. The main contractor, therefore, requires the subcontractor to put the work right even though it may involve great cost and even though the work may never in the event suffer a functional failure. The subcontractor says that this is totally unreasonable and refuses to do it, so the main contractor refuses to hand over monies otherwise due to the subcontractor, saying that unless the subcontractor rectifies the departure or provides an adequate indemnity or guarantee against future failure, the main contractor will put the work right and charge the subcontractor for it, even though the employer does not require this to be done. Can the main contractor adopt this course of action?

This type of scenario came before the Court of Appeal in the case of *Imodco Ltd* v. *Wimpey Major Products Ltd and Taylor Woodrow International Ltd.*

Imodco Ltd v. *Wimpey Major Products Ltd and Taylor Woodrow International Ltd* (1987) 40 BLR 1, CA

The facts

Imodco Ltd were the subcontractors. Wimpey and Taylor Woodrow were a joint venture contracted to the Property Services Agency (PSA) of the Department of the Environment to carry out extensive civil engineering works in the Falkland Islands, including the laying of pipelines from off-shore to the land. The joint venture engaged Imodco as subcontractors to carry out this aspect of the work.

In carrying out the work Imodco were required to install a single point mooring buoy and three submarine pipelines. The pipelines were to be laid parallel but separate from one another across the sea bed. In error, and in breach of the specification, one of the pipelines was laid underneath another. It then re-crossed underneath again to finish in its correct position. The PSA were concerned that this increased the risks of the failure of the lower line because of the weight of the upper line on top of it. Imodco provided certain calculations and gave certain assurances which satisfied the PSA that the risk of future failure of the pipeline was not great and they were accordingly prepared to issue a certificate of practical completion to the joint venture, and did so. However, in so doing they expressly reserved their rights against the joint venture should any failure occur.

The terms of main contract and subcontract are only relevant in so far as they both contained clauses making the main contractor and subcontractor respectively liable for defective work and requiring them to comply with the specification.

The joint venture were not happy and withheld money from Imodco as they were not prepared to take the risk that at some future date they might have to rectify a failure, only to find that Imodco were no longer good for the cost of rectification. They required Imodco either to provide a satisfactory guarantee from an independent and reliable source to cover the risk of future failure, or to rectify their departure from the specification. The joint venture withheld in total in the region of £1m which would otherwise have been due to Imodco. The estimated cost of rectification was £1.7m.

Imodco issued a writ and sought summary judgment under Order 14, or alternatively an interim payment under Order 29 of the Rules of the Supreme Court. The joint venture sought a stay of the court proceedings in order that the dispute might be referred to arbitration under section 4 of the Arbitration Act 1950.

Held

The judge at first instance refused to order summary judgment but ordered the joint venture to pay to Imodco £600,000 by way of an interim payment. He also stayed the proceedings to arbitration.

Imodco appealed claiming that they should have been given summary judgment under Order 14 on the basis that there was no arguable defence to their claim. However, they accepted that if there was a dispute, this should be stayed to arbitration. The joint venture appealed against the interim payment order on two grounds:

(1) that there was no jurisdiction for the court, when it intends to grant a stay under section 4 of the Arbitration Act, to first make an order for interim payment under Order 29 of the Rules of the Supreme Court;

(2) that even if the court had jurisdiction to order interim payment in these circumstances, it should not have ordered an interim payment on the facts of this case.

An affidavit had been lodged on behalf of the joint venture confirming its intention to carry out the rectification work itself if Imodco were unwilling either to do so or to provide a guarantee instead.

The Court of Appeal upheld the judge's decision not to award summary judgment under Order 14. There was an admitted departure from the specification, which was a breach of the subcontract, and therefore the joint venture had an arguable set-off and defence.

The appeal in relation to the interim payment ordered by the judge required the Court of Appeal to deal with the two issues of jurisdiction and evidence.

In relation to the jurisdiction issue, Order 29 Rule 10 provided that the court can order a defendant to make an interim payment. Rule 12 then stated that the court had to be satisfied on the hearing of the application that if the action were to proceed to trial, the plaintiff would obtain judgment against the defendant for a substantial sum of money. Rule 13 provided that the court could, if it wished, require the interim payment to be paid into court. Rule 14 empowered the court to give directions as to the further conduct of the action. Rule 17 provided for an order adjusting the interim payment if, on final judgment, it proved to be too great.

It was argued on behalf of the joint venture that the discretion and power to order interim payments could only apply where the action remained in court so that the rules of Order 29, referred to above, could operate if appropriate. If the action were stayed to arbitration, none of these rules could operate and indeed, the discretion could only be exercised if the court formed the view that if the action proceeded to trial the plaintiff would obtain judgment for a substantial sum. The court here, in staying the proceedings knew that the action could not proceed to trial.

Despite this attractive argument on the part of the joint venture, the Court of Appeal nevertheless unanimously held that there was jurisdiction in these circumstances to order an interim payment. Order 29 required that at the time the court was asked to exercise its discretion to order an interim payment, it must take into account what would be the likely outcome if the action did proceed to trial. The court could do this even though they knew that it would not proceed to trial because the proceedings were to be stayed to arbitration. It was a hypothetical exercise.

The evidence point raised the question of whether, if there was jurisdiction, the facts of the case as presented to the judge at first instance in the respective affidavits of the parties, warranted the exercise by him of his discretion to order an interim payment? The majority of the Court of Appeal held that the judge had erred in the exercise of his discretion. The judge at first instance had held, despite clear affidavit evidence to the contrary on behalf of the joint venture, that it was difficult to accept that the joint venture would carry out the works and the judge did not consider that it would have been reasonable for them to have done so having regard to the attitude taken by the PSA. The judge further decided that it was unlikely that the joint venture would in all the circumstances recover anything like the sum which they maintained was the cost of redoing the work. It was likely, said the judge, that even at a trial the joint venture would not recover more than the assessment of the possibility that, contrary to the position as it appeared at the time, they might find themselves faced at some future time with a claim by the PSA if things turned out badly.

The majority of the Court of Appeal decided that there must at least be a strong possibility that, on a full hearing of the action, the appropriate measure of damages might well be the actual cost of repair provided the joint venture had a genuine intention of carrying out the work. In all the circumstances it was not unreasonable for the joint venture to elect to carry out the rectification work, even though the pipeline might operate satisfactorily throughout its anticipated life. This being so, the cost of rectification was arguably the appropriate measure of damages. Further, the joint venture had gone so far as to agree not to carry out the rectification work provided that Imodco supplied a suitable guarantee or bond.

In the minority, Caulfield J, sitting as a judge in the Court of Appeal, took the view that the judge's exercise of his discretion to order an interim payment should not be overturned. In so doing he said:

'The breaches by the plaintiffs were admitted but there can be no set-off without proof of damage. It is certain that at the time of the hearing (i.e. the application for an interim payment) no actual damage had been suffered.'

Since the *Imodco* decision, the House of Lords has considered further the position in relation to the measure of damages, in *Forsyth Stephen* v. *Ruxley Electronics and Construction Co Ltd* (1995), but it is not considered that the Ruxley decision undermines the approach adopted by the Court of Appeal in *Imodco* in relation to the evidence point.

Further, the *Imodco* decision was concerned with the position under the Arbitration Act 1950. Bearing in mind, however, that the Arbitration Act 1996 was intended to restate and improve the law relating to arbitration, it possibly remains open to the court to exercise the jurisdiction to grant interim payments despite the existence of an arbitration agreement. Such an approach is not inconsistent with the attitude of the court as expressed in *Halki Shipping* v. *Sopex Oil* (mentioned in section 6.2.1) which was concerned with whether the court should inquire into the genuineness of a dispute, rather than whether the court should grant interim relief. Interestingly, in *Van Uden Maritime* v. *Kommandit Gesellschaft in Firma Deco-Line* (1998), the European Court of Justice declared that under article 24 of the Brussels convention a court can grant provisional and protective measures despite an agreement to refer disputes to arbitration. It is not thought that Part 25 of the CPR has changed the position as set out in *Imodco*.

Smallman Construction Ltd v. Redpath Dorman Long Ltd (1988) 47 BLR 15

This litigation arose out of Phase 4 of the Broadgate Development, involving a dispute between a steelwork subcontractor and the main contractor. Redpath Dorman Long (RDL) were the main contractor and Smallman were the subcontractor. Smallman were to erect steelwork fabricated and supplied by RDL. The subcontract was the Blue Form of domestic subcontract for use with JCT 63. It contained the usual express set-off clauses and machinery.

The facts

Smallman claimed that interim payments had not been made properly by RDL and eventually issued a writ in April 1988 claiming £579,000 plus interest. The claim was made up of alleged unpaid sums spread over all 15 interim applications. The gross value of the work based on the interim applications was £1.2m. After deduction of a 5% retention and payments already made of £571,000, the balance claimed was £579,000.

Initially RDL stated that nothing was due because the gross valuation of the Smallman work was £720,000, if 5% was deducted from this together with £571,000 already paid and £113,000 which RDL claimed to be entitled to set off in respect of delay and disruption claims, then nothing further was due. However, a second line of defence was developed, that in so far as Smallman were entitled to recover anything, RDL were entitled to set off a like amount in counterclaim. This counterclaim was for £389,000, although this included the £113,000 already set off in the first line of defence. Smallman decided to proceed by Order 29 for interim payment rather than Order 14 for summary judgment.

Held

The judge at first instance made an order in favour of Smallman for a payment on account of £250,000. RDL appealed to the Court of Appeal. Order 29 rule 12(c) provided that if on the hearing of an application the court is satisfied that if the action proceeded to trial the plaintiff would obtain judgment against the defendant for a substantial sum of money, the court may order the defendant to make an interim payment of such amount as it thinks just after taking into account any set-off, cross-claim or counterclaim on which the defendant may be entitled to rely. The same proviso applies in Part 25.7(5)(b) of the new CPR.

The Court of Appeal held that the alleged set-off of £113,000 was not the subject of an adequate notice under the set-off provisions and was therefore unsustainable. With regard to the rest of the counterclaim, inadequate details had been provided in accordance with the requirements of clause 15(3) of the contract. In addition, no loss and expense had actually been incurred in relation to this element of the claim. Smallman must, therefore, said the court, be entitled to judgment on a claim for £113,000, which was alleged to be a pure set-off claim but which did not comply with the set-off machinery. As to the balance, this concerned the alleged value of the work done and fell into a different category.

However, the whole of Smallman's claim was under Order 29 rather than Order 14 and the provisions of Order 29 required the court to be satisfied that if the action proceeded to trial the plaintiff would obtain judgment against the defendant for a substantial sum of money, a requirement carried through to the new CPR. It was impossible to see how the court could be satisfied on this. In order to be satisfied under Order 29 Rule 12(c) the court would have to be satisfied that the plaintiff would establish a substantial amount as due after a counterclaim had been heard and dealt with at trial. The counterclaim had to be taken into account as part of the process of operating the Order 29 provisions.

The Court of Appeal hinted that, if an Order 14 application for summary judgment were made in respect of the £113,000, this might be successful.

In further proceedings under Order 14, Judge Bowsher QC gave Order 14 summary judgment for £113,000. In addition he refused to stay execution of judgment. The judge refused to exercise his discretion to allow a stay, even though Smallman's business was being conducted by administrators, on the basis that the ability of the subcontractor to repay the judgment money in the event of a successful counterclaim was very low in the list of priorities.

The Court of Appeal in its decision in this case confirmed that as a result of the Court of Appeal decision in *Shanning International Ltd* v. *George Wimpey International Ltd* (1988), in any Order 29 application for interim payment, the court had to be satisfied that at a full trial the plaintiff would be entitled to a substantial sum of money after taking into account any arguable set off or counterclaim, including an independent counterclaim. This decision does somewhat restrict what had been thought previously to be the scope of the Order 29 remedy and, although there may be some scope for argument about what is meant by 'relevant' counterclaim, the wording of Part 25.7(5)(b) of the new CPR does nothing to alter this.

However, in *A.G. Machin Design & Building Contractors* v. *Mr and Mrs Long* (1992) the Court of Appeal had the opportunity to consider the test to be applied when ordering an interim payment. The application arose in unusual circumstances – after the trial of preliminary issues. The judge assumed that £60,000 (the balance of the sums certified) was due; on top of this sum was a claim for £38,000 for extras which the judge was prepared to balance against the employers' defects claim, each cancelling the other out. Against the £60,000 balance of the certified sum the judge was prepared to deduct £30,000 paid by the employers direct to subcontractors. The judge then made an interim payment order in respect of the balance of £30,000. The employers appealed.

The Court of Appeal found that the trial judge had been right to approach the matter in two stages by deciding first whether he was satisfied that, if the action proceeded to trial, the contractor would obtain judgment for a substantial sum; and secondly whether the court, in its discretion, should order an interim payment and if so in what amount.

The test to be applied was whether the court was satisfied, on the balance of probabilities but to a high standard, that the contractor would succeed in his claim for extras to an extent which at least equalled the cost of the defects counterclaim. This decision was probably confined to its own special facts, i.e. during the trial of the preliminary issues the judge had the opportunity of hearing evidence which had enabled him to conclude that there should be no deduction from the £30,000 balance he had arrived at in the contractor's favour. The Court of Appeal found that in the circumstances of the case it was the proper approach. Nonetheless the principles applied in *Shanning International Ltd* v. *George Wimpey Ltd* represent the approach that will generally be adopted by the court.

In the context of a claim for sums due in circumstances where the paying party has failed to follow contractual set-off requirements, the unpaid recipient, if seeking early payment, may use adjudication where it is available, or the Part 24 procedure. Part 25 will not assist, as in deciding whether to order an interim payment the court has to contemplate the likelihood of a substantial payment if the case went to a full trial, and in doing so must give due weight to the actual merits of any set-off, rather than ascertain simply whether the contractual requirements for an immediate set-off have been met. This is confirmed in Rule 25.7(5)(b).

The scope for the use of interim payment applications under Part 25 of the CPR is not great as the discussion of the cases indicates. For construction contracts adjudication will provide a remedy similar to that under Part 25, i.e. a decision will be given that is binding (and shall be complied with) until the dispute is finally determined by court proceedings, arbitration or agreement. Adjudicators are obliged to act impartially and to reach their decisions '... in accordance with the applicable law ...' (see the Scheme, rule 12(a)). Whether this obliges them to adopt principles analogous to those applicable to the court under Part 25 remains to be seen.

6.4   The future

As mentioned in Chapter 1 for construction contracts adjudication will become the remedy of first resort, with Part 24 for the most part used as the enforcement process for adjudicators' decisions. For non-construction contracts Part 24 will continue as the pre-eminent summary enforcement procedure for those disputes that are not subject to an arbitration agreement.

While it is sometimes possible to enlist the help of the court in obtaining money quickly, without having to go through the time consuming and expensive process of a full trial, this is clearly only appropriate where money is obviously due, so care has to be taken to select the right type of case for these summary procedures.

The use of Orders 14 and 29 was considered in the case of *Crown House Engineering Ltd* v. *Amec Products Ltd* (1989). In this case Bingham LJ said:

'The high cost of litigation, and the premium on holding cash when interest rates are high, greatly increase the attractiveness to commercial plaintiffs of procedural short-cuts such as are provided by Order 14 and Order 29 rule 12. A technical knock-out in the first round is much more advantageous than a win on points after fifteen rounds. So plaintiffs are understandably tempted to seek summary judgment or interim payment in cases for which these procedures were never intended. This is a tendency which the courts have found it necessary to discourage: *Home and Overseas Insurance Co. Ltd* v. *Mentor Insurance Co (UK) Ltd* (1989) and *British and Commonwealth Holdings plc* v. *Quadrex Holdings Inc* (1989).

These cases emphasise that Order 14 is for clear cases, that is, cases in which there is no serious material factual dispute and, if a legal issue, then no more than a crisp legal question as well decided summarily as otherwise. Order 29 rule 12 enables the court to order payment to a plaintiff to the extent that a claim, although not actually admitted, can scarcely be effectively denied. The procedure is entirely inappropriate where the plaintiff's entitlement to recover any sum is the subject of any serious dispute, whether of law or fact. This is not to say in either case that a defendant with no or no more than a partial defence can cheat a plaintiff of his just desserts by producing hefty affidavits and voluminous exhibits to create an illusion of complexity where none exists. Where the point at issue is at heart a short one, the court will recognise the fact and act accordingly no matter how bulky its outer garments. But it does mean that where there are substantial issues of genuine complexity the parties should prepare for trial (perhaps, as here, with trial of preliminary issues) rather than dissipate their energy and resources on deceptively attractive short-cuts.'

In *John Mowlem & Co Plc* v. *Carlton Gate Development Co Ltd* (1990), Judge Bowsher QC followed this approach and declined to decide issues that he considered capable of being decided under the Order 14 procedure, referring all issues to arbitration.

Further, Bingham LJ's strong judicial warning against the inappropriate use of the Order 14 procedure is likely to assume even greater importance under the new Civil Procedure Rules which reform the Order 14 procedure by empowering the court to grant summary judgment where 'that defendant has no real prospect of successfully defending the claim or issue'. Such a rule is likely to embolden claimants to seek summary judgment. Indeed it may similarly encourage defendants to adopt the corresponding procedure where it is believed '... that claimant has no real prospect of success on the claim or issue'. However if, as is considered likely, the court continues to refuse summary judgment where there is a serious factual or legal dispute, the new rules will not greatly change matters. Great caution should be exercised before such applications are made, and then they should be reserved for short crisp issues.

APPENDIX 1

HOUSING GRANTS, CONSTRUCTION AND REGENERATION ACT 1996

PART II
CONSTRUCTION CONTRACTS
Introductory provisions

104.–(1) In this Part a "construction contract" means an agreement with a person for any of the following– Construction contracts.

    (a) the carrying out of construction operations;

    (b) arranging for the carrying out of construction operations by others, whether under sub-contract to him or otherwise;

    (c) providing his own labour, or the labour of others, for the carrying out of construction operations.

(2) References in this Part to a construction contract include an agreement– 1996 c. 18.

    (a) to do architectural, design, or surveying work, or

    (b) to provide advice on building, engineering, interior or exterior decoration or on the laying-out of landscape,

in relation to construction operations.

(3) References in this Part to a construction contract do not include a contract of employment (within the meaning of the Employment Rights Act 1996).

(4) The Secretary of State may by order add to, amend or repeal any of the provisions of subsection (1), (2) or (3) as to the agreements which are construction contracts for the purposes of this Part or are to be taken or not to be taken as included in references to such contracts.

No such order shall be made unless a draft of it has been laid before and approved by a resolution of each House of Parliament.

(5) Where an agreement relates to construction operations and other matters, this Part applies to it only so far as it relates to construction operations.

An agreement relates to construction operations so far as it makes provision of any kind within subsection (1) or (2).

(6) This Part applies only to construction contracts which–
   (a) are entered into after the commencement of this Part, and
   (b) relate to the carrying out of construction operations in England, Wales or Scotland.

(7) This Part applies whether or not the law of England and Wales or Scotland is otherwise the applicable law in relation to the contract.

Meaning of "construction operations".

105.–( ) In this Part "construction operations" means, subject as follows, operations of any of the following descriptions–
   (a) construction, alteration, repair, maintenance, extension, demolition or dismantling of buildings, or structures forming, or to form, part of the land (whether permanent or not).
   (b) construction, alteration, repair, maintenance, extension, demolition or dismantling of any works forming, or to form, part of the land, including (without prejudice to the foregoing) walls, roadworks, power-lines, telecommunication apparatus, aircraft runways, docks and harbours, railways, inland waterways, pipe-lines, reservoirs, water-mains, wells, sewers, industrial plant and installations for purposes of land drainage, coast protection or defence;
   (c) installation in any building or structure of fittings forming part of the land, including (without prejudice to the foregoing) systems of heating, lighting, air-conditioning, ventilation, power supply, drainage, sanitation, water supply or fire protection, or security or communications systems;
   (d) external or internal cleaning of buildings and structures, so far as carried out in the course of their construction, alteration, repair, extension or restoration;
   (e) operations which form an integral part of, or are preparatory to, or are for rendering complete, such operations as are previously described in this subsection, including site clearance, earth-moving, excavation, tunnelling and boring, laying of foundations, erection, maintenance or dismantling of scaffolding, site restoration, landscaping and the provision of roadways and other access works;
   (f) painting or decorating the internal or external surfaces of any building or structure.

(2) The following operations are not construction operations within the meaning of this Part–
   (a) drilling for, or extraction of, oil or natural gas;
   (b) extraction (whether by underground or surface working) of minerals; tunnelling or boring, or construction of underground works, for this purpose;
   (c) assembly, installation or demolition of plant or machinery, or

erection or demolition of steelwork for the purposes of supporting or providing access to plant or machinery, on a site where the primary activity is–
> (i) nuclear processing, power generation, or water or effluent treatment, or
> (ii) the production, transmission, processing or bulk storage (other than warehousing) of chemicals, pharmaceuticals, oil, gas, steel or food and drink;

(d) manufacture or delivery to site of–
> (i) building or engineering components or equipment,
> (ii) materials, plant or machinery, or
> (iii) components for systems of heating, lighting, air-conditioning, ventilation, power supply, drainage, sanitation, water supply or fire protection, or for security or communications systems,

except under a contract which also provides for their installation;

(e) the making, installation and repair of artistic works, being sculptures, murals and other works which are wholly artistic in nature.

(3) The Secretary of State may by order add to, amend or repeal any of the provisions of subsection (1) or (2) as to the operations and work to be treated as construction operations for the purposes of this Part.

(4) No such order shall be made unless a draft of it has been laid before and approved by a resolution of each House of Parliament.

106.–(1) This Part does not apply–

> (a) to a construction contract with a residential occupier (see below), or

> (b) to any other description of construction contract excluded from the operation of this Part by order of the Secretary of State.

Provisions not applicable to contract with residential occupier.

(2) A construction contract with a residential occupier means a construction contract which principally relates to operations on a dwelling which one of the parties to the contract occupies, or intends to occupy, as his residence.

In this subsection "dwelling" means a dwelling-house or a flat; and for this purpose–
> "dwelling-house" does not include a building containing a flat; and
> "flat" means separate and self-contained premises constructed or adapted for use for residential purposes and forming part of a building from some other part of which the premises are divided horizontally.

(3) The Secretary of State may by order amend subsection (2).

(4) No order under this section shall be made unless a draft of it has been laid before and approved by a resolution of each House of Parliament.

Provisions applicable only to agreements in writing.

107.–(1) The provisions of this Part apply only where the construction contract is in writing, and any other agreement between the parties as to any matter is effective for the purposes of this Part only if in writing.

The expressions "agreement", "agree" and "agreed" shall be construed accordingly.

(2) There is an agreement in writing–
  (a) if the agreement is made in writing (whether or not it is signed by the parties),
  (b) if the agreement is made by exchange of communications in writing, or
  (c) if the agreement is evidenced in writing.

(3) Where parties agree otherwise than in writing by reference to terms which are in writing, they make an agreement in writing.

(4) An agreement is evidenced in writing if an agreement made otherwise than in writing is recorded by one of the parties, or by a third party, with the authority of the parties to the agreement.

(5) An exchange of written submissions in adjudication proceedings, or in arbitral or legal proceedings in which the existence of an agreement otherwise than in writing is alleged by one party against another party and not denied by the other party in his response constitutes as between those parties an agreement in writing to the effect alleged.

(6) References in this Part to anything being written or in writing include its being recorded by any means.

Adjudication

Right to refer disputes to adjudication.

108.–(1) A party to a construction contract has the right to refer a dispute arising under the contract for adjudication under a procedure complying with this section.

For this purpose "dispute" includes any difference.

(2) The contract shall–
  (a) enable a party to give notice at any time of his intention to refer a dispute to adjudication;
  (b) provide a timetable with the object of securing the appointment of the adjudicator and referral of the dispute to him within 7 days of such notice;
  (c) require the adjudicator to reach a decision within 28 days of

referral or such longer period as is agreed by the parties after the dispute has been referred;

(d) allow the adjudicator to extend the period of 28 days by up to 14 days, with the consent of the party by whom the dispute was referred;

(e) impose a duty on the adjudicator to act impartially; and

(f) enable the adjudicator to take the initiative in ascertaining the facts and the law.

(3) The contract shall provide that the decision of the adjudicator is binding until the dispute is finally determined by legal proceedings, by arbitration (if the contract provides for arbitration or the parties otherwise agree to arbitration) or by agreement.

The parties may agree to accept the decision of the adjudicator as finally determining the dispute.

(4) The contract shall also provide that the adjudicator is not liable for anything done or omitted in the discharge or purported discharge of his functions as adjudicator unless the act or omission is in bad faith, and that any employee or agent of the adjudicator is similarly protected from liability.

(5) If the contract does not comply with the requirements of subsections (1) to (4), the adjudication provisions of the Scheme for Construction Contracts apply.

(6) For England and Wales, the Scheme may apply the provisions of the Arbitration Act 1996 with such adaptations and modifications as appear to the Minister making the scheme to be appropriate.   1996 c. 23.

For Scotland, the Scheme may include provision conferring powers on courts in relation to adjudication and provision relating to the enforcement of the adjudicator's decision.

Payment

109.–(1) A party to a construction contract is entitled to payment by instalments, stage payments or other periodic payments for any work under the contract unless–   **Entitlement to stage payments.**

(a) it is specified in the contract that the duration of the work is to be less than 45 days, or

(b) it is agreed between the parties that the duration of the work is estimated to be less than 45 days.

(2) The parties are free to agree the amounts of the payments and the intervals at which, or circumstances in which, they become due.

(3) In the absence of such agreement, the relevant provisions of the Scheme for Construction Contracts apply.

(4) References in the following sections to a payment under the contract include a payment by virtue of this section.

Dates for payment.

110.–(1) Every construction contract shall–
  (a) provide an adequate mechanism for determining what payments become due under the contract, and when, and
  (b) provide for a final date for payment in relation to any sum which becomes due.
The parties are free to agree how long the period is to be between the date on which a sum becomes due and the final date for payment.

(2) Every construction contract shall provide for the giving of notice by a party not later than five days after the date on which a payment becomes due from him under the contract, or would have become due if–
  (a) the other party had carried out his obligations under the contract, and
  (b) no set-off or abatement was permitted by reference to any sum claimed to be due under one or more other contracts,
specifying the amount (if any) of the payment made or proposed to be made, and the basis on which that amount was calculated.

(3) If or to the extent that a contract does not contain such provision as is mentioned in subsection (1) or (2), the relevant provisions of the Scheme for Construction Contracts apply.

Notice of intention to withhold payment.

111.–(1) A party to a construction contract may not withhold payment after the final date for payment of a sum due under the contract unless he has given an effective notice of intention to withhold payment.
The notice mentioned in section 110(2) may suffice as a notice of intention to withhold payment if it complies with the requirements of this section.

(2) To be effective such a notice must specify–
  (a) the amount proposed to be withheld and the ground for withholding payment, or
  (b) if there is more than one ground, each ground and the amount attributable to it,
and must be given not later than the prescribed period before the final date for payment.

(3) The parties are free to agree what that prescribed period is to be.
In the absence of such agreement, the period shall be that provided by the Scheme for Construction Contracts.

(4) Where an effective notice of intention to withhold payment is given, but on the matter being referred to adjudication it is decided that

the whole or part of the amount should be paid, the decision shall be construed as requiring payment not later than–
  (a) seven days from the date of the decision, or
  (b) the date which apart from the notice would have been the final date for payment,
whichever is the later

112.–(1) Where a sum due under a construction contract is not paid in full by the final date for payment and no effective notice to withhold payment has been given, the person to whom the sum is due has the right (without prejudice to any other right or remedy) to suspend performance of his obligations under the contract to the party by whom payment ought to have been made ("the party in default").

Right to suspend performance for non-payment.

(2) The right may not be exercised without first giving to the party in default at least seven days' notice of intention to suspend performance, stating the ground or grounds on which it is intended to suspend performance.

(3) The right to suspend performance ceases when the party in default makes payment in full of the amount due.

(4) Any period during which performance is suspended in pursuance of the right conferred by this section shall be disregarded in computing for the purposes of any contractual time limit the time taken, by the party exercising the right or by a third party, to complete any work directly or indirectly affected by the exercise of the right.
  Where the contractual time limit is set by reference to a date rather than a period, the date shall be adjusted accordingly.

113.–(1) A provision making payment under a construction contract conditional on the payer receiving payment from a third person is ineffective, unless that third person, or any other person payment by whom is under the contract (directly or indirectly) a condition of payment by that third person, is insolvent.

Prohibition of conditional payment provisions.

(2) For the purposes of this section a company becomes insolvent–
  (a) on the making of an administration order against it under Part II of the Insolvency Act 1986,
  (b) on the appointment of an administrative receiver or a receiver or manager of its property under Chapter I of Part III of that Act, or the appointment of a receiver under Chapter II of that Part,
  (c) on the passing of a resolution for voluntary winding-up without a declaration of solvency under section 89 of that Act, or

1986 c. 45.

(d) on the making of a winding-up order under Part IV or V of that Act.

(3) For the purposes of the section a partnership becomes insolvent–
　(a) on the making of a winding-up order against it under any provision of the Insolvency Act 1986 as applied by an order under section 420 of that Act, or
　(b) when sequestration is awarded on the estate of the partnership under section 12 of the Bankruptcy (Scotland) Act 1985 or the partnership grants a trust deed for its creditors.

1985 c. 66.

(4) For the purposes of this section an individual becomes insolvent–
　(a) on the making of a bankruptcy order against him under Part IX of the Insolvency Act 1986, or
　(b) on the sequestration of his estate under the Bankruptcy (Scotland) Act 1985 or when he grants a trust deed for his creditors.

1986 c. 45.

(5) A company, partnership or individual shall also be treated as insolvent on the occurrence of any event corresponding to those specified in subsection (2), (3) or (4) under the law of Northern Ireland or of a country outside the United Kingdom.

(6) Where a provision is rendered ineffective by subsection (1), the parties are free to agree other terms for payment.
In the absence of such agreement, the relevant provisions of the Scheme for Construction Contracts apply.

Supplementary provisions

The Scheme for Construction Contracts.

114.–(1) The Minister shall by regulations make a scheme ("the Scheme for Construction Contracts") containing provision about the matters referred to in the preceding provisions of this Part.

(2) Before making any regulations under this section the Minister shall consult such persons as he thinks fit.

(3) In this section "the Minister" means–
　(a) for England and Wales, the Secretary of State, and
　(b) for Scotland, the Lord Advocate.

(4) Where any provisions of the Scheme for Construction Contracts apply by virtue of this Part in default of contractual provision agreed by the parties, they have effect as implied terms of the contract concerned.

(5) Regulations under this section shall not be made unless a draft of them has been approved by resolution of each House of Parliament.

Service of notices, &c.

115.–(1) The parties are free to agree on the manner of service of any notice or other document required or authorised to be served in pursuance of the construction contract or for any of the purposes of this Part.

(2) If or to the extent that there is no such agreement the following provisions apply.

(3) A notice or other document may be served on a person by any effective means.

(4) If a notice or other document is addressed, pre-paid and delivered by post–
    (a) to the addressee's last known principal residence or, if he is or has been carrying on a trade, profession or business, his last known principal business address, or
    (b) where the addressee is a body corporate, to the body's registered or principal office,
it shall be treated as effectively served.

(5) This section does not apply to the service of documents for the purposes of legal proceedings, for which provision is made by rules of court.

(6) References in this Part to a notice or other document include any form of communication in writing and references to service shall be construed accordingly.

116.–(1) For the purposes of this Part periods of time shall be reckoned as follows.

Reckoning periods of time.

(2) Where an act is required to be done within a specified period after or from a specified date, the period begins immediately after that date.

(3) Where the period would include Christmas Day, Good Friday or a day which under the Banking and Financial Dealings Act 1971 is a bank holiday in England and Wales or, as the case may be, in Scotland, that day shall be excluded.

1971 c. 80.

117.–(1) This Part applies to a construction contract entered into by or on behalf of the Crown otherwise than by or on behalf of Her Majesty in her private capacity.

Crown application.

(2) This Part applies to a construction contract entered into on behalf of the Duchy of Cornwall notwithstanding any Crown interest.

(3) Where a construction contract is entered into by or on behalf of Her Majesty in right of the Duchy of Lancaster, Her Majesty shall be represented, for the purposes of any adjudication or other proceedings arising out of the contract by virtue of this Part, by the Chancellor of the Duchy or such person as he may appoint.

(4) Where a construction contract is entered into on behalf of the Duchy of Cornwall, the Duke of Cornwall or the possessor for the time being of the Duchy shall be represented, for the purposes of any adjudication or other proceedings arising out of the contract by virtue of this Part, by such person as he may appoint.

APPENDIX 2
THE SCHEME FOR CONSTRUCTION CONTRACTS (ENGLAND AND WALES) REGULATIONS 1998

Statutory Instrument 1998 No. 649

The Secretary of State, in exercise of the powers conferred on him by sections 108(6), 114 and 146(1) and (2) of the Housing Grants, Construction and Regeneration Act 1996, and of all other powers enabling him in that behalf, having consulted such persons as he thinks fit, and draft Regulations having been approved by both Houses of Parliament, hereby makes the following Regulations:

Citation, commencement, extent and interpretation

1.–(1) These Regulations may be cited as the Scheme for Construction Contracts (England and Wales) Regulations 1998 and shall come into force at the end of the period of 8 weeks beginning with the day on which they are made (the "commencement date").

(2) These Regulations shall extend only to England and Wales.

(3) In these Regulations, "the Act" means the Housing Grants, Construction and Regeneration Act 1996.

The Scheme for Construction Contracts

2. Where a construction contract does not comply with the requirements of section 108(1) to (4) of the Act, the adjudication provisions in Part I of the Schedule to these Regulations shall apply.

3. Where–
    (a) the parties to a construction contract are unable to reach agreement for the purposes mentioned respectively in sections 109, 111 and 113 of the Act, or
    (b) a construction contract does not make provision as required by section 110 of the Act,

117

the relevant provisions in Part II of the Schedule to these Regulations shall apply.

4. The provisions in the Schedule to these Regulations shall be the Scheme for Construction Contracts for the purposes of section 114 of the Act.

SCHEDULE   Regulations 2, 3 and 4

THE SCHEME FOR CONSTRUCTION CONTRACTS
PART I–ADJUDICATION

Notice of Intention to seek Adjudication

1.-(1) Any party to a construction contract (the "referring party") may give written notice (the "notice of adjudication") of his intention to refer any dispute arising under the contract, to adjudication.

(2) The notice of adjudication shall be given to every other party to the contract.

(3) The notice of adjudication shall set out briefly–
   (a) the nature and a brief description of the dispute and of the parties involved,
   (b) details of where and when the dispute had arisen,
   (c) the nature of the redress which is sought, and
   (d) the names and addresses of the parties to the contract (including, where appropriate, the addresses which the parties have specified for the giving of notices).

2.-(1) Following the giving of a notice of adjudication and subject to any agreement between the parties to the dispute as to who shall act as adjudicator–
   (a) the referring party shall request the person (if any) specified in the contract to act as adjudicator, or
   (b) if no person is named in the contract or the person named has already indicated that he is unwilling or unable to act, and the contract provides for a specified nominating body to select a person, the referring party shall request the nominating body named in the contract to select a person to act as adjudicator, or
   (c) where neither paragraph (a) nor (b) above applies, or where the person referred to in (a) has already indicated that he is unwilling or unable to act and (b) does not apply, the referring party shall request an adjudicator nominating body to select a person to act as adjudicator.

(2) A person requested to act as adjudicator in accordance with the provisions of paragraph (1) shall indicate whether or not he is willing to act within two days of receiving the request.

(3) In this paragraph, and in paragraphs 5 and 6 below, an "adjudicator nominating body" shall mean a body (not being a natural person and not being a party to the dispute) which holds itself out publicly as a body which will select an adjudicator when requested to do so by a referring party.

3. The request referred to in paragraphs 2, 5 and 6 shall be accompanied by a copy of the notice of adjudication.

4. Any person requested or selected to act as adjudicator in accordance with paragraphs 2, 5 or 6 shall be a natural person acting in his personal capacity. A person requested or selected to act as an adjudicator shall not be an employee of any of the parties to the dispute and shall declare any interest, financial or otherwise, in any matter relating to the dispute.

5.-(1) The nominating body referred to in paragraphs 2(1)(b) and 6(1)(b) or the adjudicator nominating body referred to in paragraphs 2(1)(c), 5(2)(b) and 6(1)(c) must communicate the selection of an adjudicator to the referring party within five days of receiving a request to do so.

(2) Where the nominating body or the adjudicator nominating body fails to comply with paragraph (1), the referring party may–
    (a) agree with the other party to the dispute to request a specified person to act as adjudicator, or
    (b) request any other adjudicator nominating body to select a person to act as adjudicator.

(3) The person requested to act as adjudicator in accordance with the provisions of paragraphs (1) or (2) shall indicate whether or not he is willing to act within two days of receiving the request.

6.-(1) Where an adjudicator who is named in the contract indicates to the parties that he is unable or unwilling to act, or where he fails to respond in accordance with paragraph 2(2), the referring party may–
    (a) request another person (if any) specified in the contract to act as adjudicator, or
    (b) request the nominating body (if any) referred to in the contract to select a person to act as adjudicator, or
    (c) request any other adjudicator nominating body to select a person to act as adjudicator.

(2) The person requested to act in accordance with the provisions of paragraph (1) shall indicate whether or not he is willing to act within two days of receiving the request.

7.–(1) Where an adjudicator has been selected in accordance with paragraphs 2, 5 or 6, the referring party shall, not later than seven days from the date of the notice of adjudication, refer the dispute in writing (the "referral notice") to the adjudicator.

(2) A referral notice shall be accompanied by copies of, or relevant extracts from, the construction contract and such other documents as the referring party intends to rely upon.

(3) The referring party shall, at the same time as he sends to the adjudicator the documents referred to in paragraphs (1) and (2), send copies of those documents to every other party to the dispute.

8.–(1) The adjudicator may, with the consent of all the parties to those disputes, adjudicate at the same time on more than one dispute under the same contract.

(2) The adjudicator may, with the consent of all the parties to those disputes, adjudicate at the same time on related disputes under different contracts, whether or not one or more of those parties is a party to those disputes.

(3) All the parties in paragraphs (1) and (2) respectively may agree to extend the period within which the adjudicator may reach a decision in relation to all or any of these disputes.

(4) Where an adjudicator ceases to act because a dispute is to be adjudicated on by another person in terms of this paragraph, that adjudicator's fees and expenses shall be determined in accordance with paragraph 25.

9.–(1) An adjudicator may resign at any time on giving notice in writing to the parties to the dispute.

(2) An adjudicator must resign where the dispute is the same or substantially the same as one which has previously been referred to adjudication, and a decision has been taken in that adjudication.

(3) Where an adjudicator ceases to act under paragraph 9(1)–
  (a) the referring party may serve a fresh notice under paragraph 1 and shall request an adjudicator to act in accordance with paragraphs 2 to 7; and
  (b) if requested by the new adjudicator and insofar as it is reasonably practicable, the parties shall supply him with copies of all documents which they had made available to the previous adjudicator.

(4) Where an adjudicator resigns in the circumstances referred to in paragraph (2), or where a dispute varies significantly from the dispute referred to him in the referral notice and for that reason he is not

competent to decide it, the adjudicator shall be entitled to the payment of such reasonable amount as he may determine by way of fees and expenses reasonably incurred by him. The parties shall be jointly and severally liable for any sum which remains outstanding following the making of any determination on how the payment shall be apportioned.

10. Where any party to the dispute objects to the appointment of a particular person as adjudicator, that objection shall not invalidate the adjudicator's appointment nor any decision he may reach in accordance with paragraph 20.

11.–(1) The parties to a dispute may at any time agree to revoke the appointment of the adjudicator. The adjudicator shall be entitled to the payment of such reasonable amount as he may determine by way of fees and expenses incurred by him. The parties shall be jointly and severally liable for any sum which remains outstanding following the making of any determination on how the payment shall be apportioned.

(2) Where the revocation of the appointment of the adjudicator is due to the default or misconduct of the adjudicator, the parties shall not be liable to pay the adjudicator's fees and expenses.

Powers of the adjudicator

12. The adjudicator shall–
   (a) act impartially in carrying out his duties and shall do so in accordance with any relevant terms of the contract and shall reach his decision in accordance with the applicable law in relation to the contract; and
   (b) avoid incurring unnecessary expense.

13. The adjudicator may take the initiative in ascertaining the facts and the law necessary to determine the dispute, and shall decide on the procedure to be followed in the adjudication. In particular he may–
   (a) request any party to the contract to supply him with documents as he may reasonably require including, if he so directs, any written statement from any party to the contract supporting or supplementing the referral notice and any other documents given under paragraph 7(2),
   (b) decide the language or languages to be used in the adjudication and whether a translation of any document is to be provided and if so by whom,
   (c) meet and question any of the parties to the contract and their representatives,
   (d) subject to obtaining any necessary consent from a third party or parties, make such site visits and inspections as he considers appropriate, whether accompanied by the parties or not,

(e) subject to obtaining any necessary consent from a third party or parties, carry out any tests or experiments,

(f) obtain and consider such representations and submissions as he requires, and, provided he has notified the parties of his intention, appoint experts, assessors or legal advisers,

(g) give directions as to the timetable for the adjudication, any deadlines, or limits as to the length of written documents or oral representations to be complied with, and

(h) issue other directions relating to the conduct of the adjudication.

14. The parties shall comply with any request or direction of the adjudicator in relation to the adjudication.

15. If, without showing sufficient cause, a party fails to comply with any request, direction or timetable of the adjudicator made in accordance with his powers, fails to produce any document or written statement requested by the adjudicator, or in any other way fails to comply with a requirement under these provisions relating to the adjudication, the adjudicator may–

(a) continue the adjudication in the absence of that party or of the document or written statement requested,

(b) draw such inferences from that failure to comply as circumstances may, in the adjudicator's opinion, be justified, and

(c) make a decision on the basis of the information before him attaching such weight as he thinks fit to any evidence submitted to him outside any period he may have requested or directed.

16.–(1) Subject to any agreement between the parties to the contrary, and to the terms of paragraph (2) below, any party to the dispute may be assisted by, or represented by, such advisers or representatives (whether legally qualified or not) as he considers appropriate.

(2) Where the adjudicator is considering oral evidence or representations, a party to the dispute may not be represented by more than one person, unless the adjudicator gives directions to the contrary.

17. The adjudicator shall consider any relevant information submitted to him by any of the parties to the dispute and shall make available to them any information to be taken into account in reaching his decision.

18. The adjudicator and any party to the dispute shall not disclose to any other person any information or document provided to him in connection with the adjudication which the party supplying it has indicated is to be treated as confidential, except to the extent that it is necessary for the purposes of, or in connection with, the adjudication.

19.–(1) The adjudicator shall reach his decision not later than–

(a) twenty eight days after the date of the referral notice mentioned in paragraph 7(1), or

(b) forty two days after the date of the referral notice if the referring party so consents, or

(c) such period exceeding twenty eight days after the referral notice as the parties to the dispute may, after the giving of that notice, agree.

(2) Where the adjudicator fails, for any reason, to reach his decision in accordance with paragraph (1)

(a) any of the parties to the dispute may serve a fresh notice under paragraph 1 and shall request an adjudicator to act in accordance with paragraphs 2 to 7; and

(b) if requested by the new adjudicator and insofar as it is reasonably practicable, the parties shall supply him with copies of all documents which they had made available to the previous adjudicator.

(3) As soon as possible after he has reached a decision, the adjudicator shall deliver a copy of that decision to each of the parties to the contract.

Adjudicator's decision

20. The adjudicator shall decide the matters in dispute. He may take into account any other matters which the parties to the dispute agree should be within the scope of the adjudication or which are matters under the contract which he considers are necessarily connected with the dispute. In particular, he may–

(a) open up, revise and review any decision taken or any certificate given by any person referred to in the contract unless the contract states that the decision or certificate is final and conclusive,

(b) decide that any of the parties to the dispute is liable to make a payment under the contract (whether in sterling or some other currency) and, subject to section 111(4) of the Act, when that payment is due and the final date for payment,

(c) having regard to any term of the contract relating to the payment of interest decide the circumstances in which, and the rates at which, and the periods for which simple or compound rates of interest shall be paid.

21. In the absence of any directions by the adjudicator relating to the time for performance of his decision, the parties shall be required to comply with any decision of the adjudicator immediately on delivery of the decision to the parties in accordance with this paragraph.

22. If requested by one of the parties to the dispute, the adjudicator shall provide reasons for his decision.

Effects of the decision

23.–(1) In his decision, the adjudicator may, if he thinks fit, order any of the parties to comply peremptorily with his decision or any part of it.

(2) The decision of the adjudicator shall be binding on the parties, and they shall comply with it until the dispute is finally determined by legal proceedings, by arbitration (if the contract provides for arbitration or the parties otherwise agree to arbitration) or by agreement between the parties.

24. Section 42 of the Arbitration Act 1996 shall apply to this Scheme subject to the following modifications–
   (a) in subsection (2) for the word "tribunal" wherever it appears there shall be substituted the word "adjudicator",
   (b) in subparagraph (b) of subsection (2) for the words "arbitral proceedings" there shall be substituted the word "adjudication",
   (c) subparagraph (c) of subsection (2) shall be deleted, and
   (d) subsection (3) shall be deleted.

25. The adjudicator shall be entitled to the payment of such reasonable amount as he may determine by way of fees and expenses reasonably incurred by him. The parties shall be jointly and severally liable for any sum which remains outstanding following the making of any determination on how the payment shall be apportioned.

26. The adjudicator shall not be liable for anything done or omitted in the discharge or purported discharge of his functions as adjudicator unless the act or omission is in bad faith, and any employee or agent of the adjudicator shall be similarly protected from liability.

PART II–PAYMENT

Entitlement to and amount of stage payments

1. Where the parties to a relevant construction contract fail to agree–
   (a) the amount of any instalment or stage or periodic payment for any work under the contract, or
   (b) the intervals at which, or circumstances in which, such payments become due under that contract, or
   (c) both of the matters mentioned in sub-paragraphs (a) and (b) above,
the relevant provisions of paragraphs 2 to 4 below shall apply.

2.-(1) The amount of any payment by way of instalments or stage or periodic payments in respect of a relevant period shall be the difference between the amount determined in accordance with sub-paragraph (2) and the amount determined in accordance with sub-paragraph (3).

(2) The aggregate of the following amounts–
  (a) an amount equal to the value of any work performed in accordance with the relevant construction contract during the period from the commencement of the contract to the end of the relevant period (excluding any amount calculated in accordance with sub-paragraph (b)),
  (b) where the contract provides for payment for materials, an amount equal to the value of any materials manufactured on site or brought onto site for the purposes of the works during the period from the commencement of the contract to the end of the relevant period, and
  (c) any other amount or sum which the contract specifies shall be payable during or in respect of the period from the commencement of the contract to the end of the relevant period.

(3) The aggregate of any sums which have been paid or are due for payment by way of instalments, stage or periodic payments during the period from the commencement of the contract to the end of the relevant period.

(4) An amount calculated in accordance with this paragraph shall not exceed the difference between–
  (a) the contract price, and
  (b) the aggregate of the instalments or stage or periodic payments which have become due.

Dates for payment

3. Where the parties to a construction contract fail to provide an adequate mechanism for determining either what payments become due under the contract, or when they become due for payment, or both, the relevant provisions of paragraphs 4 to 7 shall apply.

4. Any payment of a kind mentioned in paragraph 2 above shall become due on whichever of the following dates occurs later–
  (a) the expiry of 7 days following the relevant period mentioned in paragraph 2(1) above, or
  (b) the making of a claim by the payee.

5. The final payment payable under a relevant construction contract, namely the payment of an amount equal to the difference (if any) between–

(a) the contract price, and
(b) the aggregate of any instalment or stage or periodic payments which have become due under the contract,
shall become due on the expiry of–
(a) 30 days following completion of the work, or
(b) the making of a claim by the payee,
whichever is the later.

6. Payment of the contract price under a construction contract (not being a relevant construction contract) shall become due on
(a) the expiry of 30 days following the completion of the work, or
(b) the making of a claim by the payee,
whichever is the later.

7. Any other payment under a construction contract shall become due
(a) on the expiry of 7 days following the completion of the work to which the payment relates, or
(b) the making of a claim by the payee,
whichever is the later.

Final date for payment

8.–(1) Where the parties to a construction contract fail to provide a final date for payment in relation to any sum which becomes due under a construction contract, the provisions of this paragraph shall apply.

(2) The final date for the making of any payment of a kind mentioned in paragraphs 2, 5, 6 or 7, shall be 17 days from the date that payment becomes due.

Notice specifying amount of payment

9. A party to a construction contract shall, not later than 5 days after the date on which any payment–
(a) becomes due from him, or
(b) would have become due, if–
(i) the other party had carried out his obligations under the contract, and
(ii) no set-off or abatement was permitted by reference to any sum claimed to be due under one or more other contracts,
give notice to the other party to the contract specifying the amount (if any) of the payment he has made or proposes to make, specifying to what the payment relates and the basis on which that amount is calculated.

Notice of intention to withhold payment

10. Any notice of intention to withhold payment mentioned in section 111 of the Act shall be given not later than the prescribed period, which is to say not later than 7 days before the final date for payment determined either in accordance with the construction contract, or where no such provision is made in the contract, in accordance with paragraph 8 above.

Prohibition of conditional payment provisions

11. Where a provision making payment under a construction contract conditional on the payer receiving payment from a third person is ineffective as mentioned in section 113 of the Act, and the parties have not agreed other terms for payment, the relevant provisions of–

    (a) paragraphs 2, 4, 5, 7, 8, 9 and 10 shall apply in the case of a relevant construction contract, and

    (b) paragraphs 6, 7, 8, 9 and 10 shall apply in the case of any other construction contract.

Interpretation

12. In this Part of the Scheme for Construction Contracts–

"claim by the payee" means a written notice given by the party carrying out work under a construction contract to the other party specifying the amount of any payment or payments which he considers to be due and the basis on which it is, or they are calculated;

"contract price" means the entire sum payable under the construction contract in respect of the work;

"relevant construction contract" means any construction contract other than one–

    (a) which specifies that the duration of the work is to be less than 45 days, or

    (b) in respect of which the parties agree that the duration of the work is estimated to be less than 45 days;

"relevant period" means a period which is specified in, or is calculated by reference to the construction contract or where no such period is so specified or is so calculable, a period of 28 days;

"value of work" means an amount determined in accordance with the construction contract under which the work is performed or where the contract contains no such provision, the cost of any work performed in accordance with that contract together with an amount equal to any overhead or profit included in the contract price;

"work" means any of the work or services mentioned in section 104 of the Act.

EXPLANATORY NOTE

(This note is not part of the Order)

Part II of the Housing Grants, Construction and Regeneration Act 1996 makes provision in relation to construction contracts. Section 114 empowers the Secretary of State to make the Scheme for Construction Contracts. Where a construction contract does not comply with the requirements of sections 108 to 111 (adjudication of disputes and payment provisions), and section 113 (prohibition of conditional payment provisions), the relevant provisions of the Scheme for Construction Contracts have effect.

The Scheme which is contained in the Schedule to these Regulations is in two parts. Part I provides for the selection and appointment of an adjudicator, gives powers to the adjudicator to gather and consider information, and makes provisions in respect of his decisions. Part II makes provision with respect to payments under a construction contract where either the contract fails to make provision or the parties fail to agree–

    (a) the method for calculating the amount of any instalment, stage or periodic payment,

    (b) the due date and the final date for payments to be made, and

    (c) prescribes the period within which a notice of intention to withhold payment must be given.

APPENDIX 3
THE CONSTRUCTION CONTRACTS (ENGLAND AND WALES) EXCLUSION ORDER 1998

Statutory Instrument 1998 No. 648

The Secretary of State, in exercise of the powers conferred on him by sections 106(1)(b) and 146(1) of the Housing Grants, Construction and Regeneration Act 1996 and of all other powers enabling him in that behalf, hereby makes the following Order, a draft of which has been laid before and approved by resolution of, each House of Parliament:

Citation, commencement and extent

1.-(1) This Order may be cited as the Construction Contracts Exclusion Order 1998 and shall come into force at the end of the period of 8 weeks beginning with the day on which it is made ("the commencement date").

(2) This Order shall extend to England and Wales only.

Interpretation

2. In this Order, "Part II" means Part II of the Housing Grants, Construction and Regeneration Act 1996.

Agreements under statute

3. A construction contract is excluded from the operation of Part II if it is–

  (a) an agreement under section 38 (power of highway authorities to adopt by agreement) or section 278 (agreements as to execution of works) of the Highways Act 1980;

  (b) an agreement under section 106 (planning obligations), 106A (modification or discharge of planning obligations) or 299A (Crown planning obligations) of the Town and Country Planning Act 1990;

(c) an agreement under section 104 of the Water Industry Act 1991 (agreements to adopt sewer, drain or sewage disposal works); or

(d) an externally financed, development agreement within the meaning of section 1 of the National Health Service (Private Finance) Act 1997 (powers of NHS Trusts to enter into agreements).

Private finance initiative

4..–(1) A construction contract is excluded from the operation of Part II if it is a contract entered into under the private finance initiative, within the meaning given below.

(2) A contract is entered into under the private finance initiative if all the following conditions are fulfilled–

(a) it contains a statement that it is entered into under that initiative or, as the case may be, under a project applying similar principles;

(b) the consideration due under the contract is determined at least in part by reference to one or more of the following–

(i) the standards attained in the performance of a service, the provision of which is the principal purpose or one of the principal purposes for which the building or structure is constructed;

(ii) the extent, rate or intensity of use of all or any part of the building or structure in question; or

(iii) the right to operate any facility in connection with the building or structure in question; and

(c) one of the parties to the contract is–

(i) a Minister of the Crown;

(ii) a department in respect of which appropriation accounts are required to be prepared under the Exchequer and Audit Departments Act 1866;

(iii) any other authority or body whose accounts are required to be examined and certified by or are open to the inspection of the Comptroller and Auditor General by virtue of an agreement entered into before the commencement date or by virtue of any enactment;

(iv) any authority or body listed in Schedule 4 to the National Audit Act 1983 (nationalised industries and other public authorities);

(v) a body whose accounts are subject to audit by auditors appointed by the Audit Commission;

(vi) the governing body or trustees of a voluntary school within the meaning of section 31 of the Education Act 1996 (county schools and voluntary schools), or

> (vii) a company wholly owned by any of the bodies described in paragraphs (i) to (v).

Finance agreements

5.–(1) A construction contract is excluded from the operation of Part II if it is a finance agreement, within the meaning given below.

(2) A contract is a finance agreement if it is any one of the following–
  (a) any contract of insurance;
  (b) any contract under which the principal obligations include the formation or dissolution of a company, unincorporated association or partnership;
  (c) any contract under which the principal obligations include the creation or transfer of securities or any right or interest in securities;
  (d) any contract under which the principal obligations include the lending of money;
  (e) any contract under which the principal obligations include an undertaking by a person to be responsible as surety for the debt or default of another person, including a fidelity bond, advance payment bond, retention bond or performance bond.

Development agreements

6.–(1) A construction contract is excluded from the operation of Part II if it is a development agreement, within the meaning given below.

(2) A contract is a development agreement if it includes provision for the grant or disposal of a relevant interest in the land on which take place the principal construction operations to which the contract relates.

(3) In paragraph (2) above, a relevant interest in land means–
  (a) a freehold; or
  (b) a leasehold for a period which is to expire no earlier than 12 months after the completion of the construction operations under the contract.

EXPLANATORY NOTE

(This note is not part of the Order)

Part II of the Housing Grants, Construction and Regeneration Act 1996 makes provision in relation to the terms of construction contracts. Section 106 confers power on the Secretary of State to exclude descriptions of contracts from the operation of Part II. This Order excludes contracts of four descriptions.

Article 3 excludes agreements made under specified statutory provisions dealing with highway works, planning obligations, sewage works and externally financed NHS Trust agreements. Article 4 excludes agreements entered into by specified public bodies under the private finance initiative (or a project applying similar principles). Article 5 excludes agreements which primarily relate to the financing of works. Article 6 excludes development agreements, which contain provision for the disposal of an interest in land.

APPENDIX 4
ARBITRATION ACT 1996

9.-(1) A party to an arbitration agreement against whom legal proceedings are brought (whether by way of claim or counterclaim) in respect of a matter which under the agreement is to be referred to arbitration may (upon notice to the other parties to the proceedings) apply to the court in which the proceedings have been brought to stay the proceedings so far as they concern that matter.

Stay of legal proceedings.

(2) An application may be made notwithstanding that the matter is to be referred to arbitration only after the exhaustion of other dispute resolution procedures.

(3) An application may not be made by a person before taking the appropriate procedural step (if any) to acknowledge the legal proceedings against him or after he has taken any step in those proceedings to answer the substantive claim.

(4) On an application under this section the court shall grant a stay unless satisfied that the arbitration agreement is null and void, inoperative, or incapable of being performed.

(5) If the court refuses to stay the legal proceedings, any provision that an award is a condition precedent to the bringing of legal proceedings in respect of any matter is of no effect in relation to those proceedings.

39.-(1) The parties are free to agree that the tribunal shall have power to order on a provisional basis any relief which it would have power to grant in a final award.

Power to make provisional awards.

(2) This includes, for instance, making–
  (a) a provisional order for the payment of money or the disposition of property as between the parties, or
  (b) an order to make an interim payment on account of the costs of the arbitration.

(3) Any such order shall be subject to the tribunal's final adjudication;

and the tribunal's final award, on the merits or as to costs, shall take account of any such order.

(4) Unless the parties agree to confer such power on the tribunal, the tribunal has no such power.

This does not affect its powers under section 47 (awards on different issues, &c.).

Awards on different issues, &c.

47.–(1) Unless otherwise agreed by the parties, the tribunal may make more than one award at different times on different aspects of the matters to be determined.

(2) The tribunal may, in particular, make an award relating–
    (a) to an issue affecting the whole claim, or
    (b) to a part only of the claims or cross-claims submitted to it for decision.

(3) If the tribunal does so, it shall specify in its award the issue, or the claim or part of a claim, which is the subject matter of the award.

APPENDIX 5
THE UNFAIR TERMS IN CONSUMER CONTRACTS REGULATIONS 1994

Statutory Instrument 1994 No. 3159

Citation and commencement

1. These Regulations may be cited as the Unfair Terms in Consumer Contracts Regulations 1994 and shall come into force on 1st July 1995.

Interpretation

2.–(1) In these Regulations–

'business' includes a trade or profession and the activities of any government department or local or public authority;

'the Community' means the European Economic Community and the other States in the European Economic Area;

'consumer' means a natural person who, in making a contract to which these Regulations apply, is acting for purposes which are outside his business;

'court' in relation to England and Wales and Northern Ireland means the High Court, and in relation to Scotland, the Court of Session;

'Director' means the Director General of Fair Trading;

'EEA Agreement' means the Agreement on the European Economic Area signed at Oporto on 2 May 1992 as adjusted by the protocol signed at Brussels on 17 March 1993;

'member State' shall mean a State which is a contracting party to the EEA Agreement but until the EEA Agreement comes into force in relation to Leichtenstein does not include the State of Liechtenstein;

'seller' means a person who sells goods and who, in making a contract to which these Regulations apply, is acting for purposes relating to his business; and

'supplier' means a person who supplies goods or services and who, in making a contract to which these Regulations apply, is acting for purposes relating to his business.

(2) In the application of these Regulations to Scotland for references to an 'injunction' or an 'interlocutory injunction' there shall be substituted references to an 'interdict' or 'interim interdict' respectively.

135

Terms to which these Regulations apply

3.-(1) Subject to the provisions of Schedule 1, these Regulations apply to any term in a contract concluded between a seller or supplier and a consumer where the said term has not been individually negotiated.

(2) In so far as it is in plain, intelligible language, no assessment shall be made of the fairness of any term which–

    (a) defines the main subject matter of the contract, or
    (b) concerns the adequacy of the price or remuneration, as against the goods or services sold or supplied.

(3) For the purposes of these Regulations, a term shall always be regarded as not having been individually negotiated where it has been drafted in advance and the consumer has not been able to influence the substance of the term.

(4) Notwithstanding that a specific term or certain aspects of it in a contract has been individually negotiated, these Regulations shall apply to the rest of a contract if an overall assessment of the contract indicates that it is a pre-formulated standard contract.

(5) It shall be for any seller or supplier who claims that a term was individually negotiated to show that it was.

Unfair terms

4.-(1) In these Regulations, subject to paragraphs (2) and (3) below, 'unfair term' means any term which contrary to the requirement of good faith causes a significant imbalance in the parties' rights and obligations under the contract to the detriment of the consumer.

(2) An assessment of the unfair nature of a term shall be made taking into account the nature of the goods or services for which the contract was concluded and referring, as at the time of the conclusion of the contract, to all circumstances attending the conclusion of the contract and to all the other terms of the contract or of another contract on which it is dependent.

(3) In determining whether a term satisfies the requirement of good faith, regard shall be had in particular to the matters specified in Schedule 2 to these Regulations.

(4) Schedule 3 to these Regulations contains an indicative and non-exhaustive list of the terms which may be regarded as unfair.

Consequence of inclusion of unfair terms in contracts

5.-(1) An unfair term in a contract concluded with a consumer by a seller or supplier shall not be binding on the consumer.

(2) The contract shall continue to bind the parties if it is capable of continuing in existence without the unfair term.

Construction of written contracts

6 A seller or supplier shall ensure that any written term of a contract is expressed in plain, intelligible language, and if there is doubt about the meaning of a written term, the interpretation most favourable to the consumer shall prevail.

Choice of law clauses

7. These Regulations shall apply notwithstanding any contract term which applies or purports to apply the law of a non member State, if the contract has a close connection with the territory of the member States.

Prevention of continued use of unfair terms

8.–(1) It shall be the duty of the Director to consider any complaint made to him that any contract term drawn up for general use is unfair, unless the complaint appears to the Director to be frivolous or vexatious.

(2) If having considered a complaint about any contract term pursuant to paragraph (1) above the Director considers that the contract term is unfair he may, if he considers it appropriate to do so, bring proceedings for an injunction (in which proceedings he may also apply for an interlocutory injunction) against any person appearing to him to be using or recommending use of such a term in contracts concluded with consumers.

(3) The Director may, if he considers it appropriate to do so, have regard to any undertakings given to him by or on behalf of any person as to the continued use of such a term in contracts concluded with consumers.

(4) The Director shall give reasons for his decision to apply or not to apply, as the case may be, for an injunction in relation to any complaint which these Regulations require him to consider.

(5) The court on an application by the Director may grant an injunction on such terms as it thinks fit.

(6) An injunction may relate not only to use of a particular contract term drawn up for general use but to any similar term, or a term having like effect, used or recommended for use by any party to the proceedings.

(7) The Director may arrange for the dissemination in such form and manner as he considers appropriate of such information and advice

concerning the operation of these Regulations as may appear to him to be expedient to give to the public and to all persons likely to be affected by these Regulations.

SCHEDULE 1 Regulation 3(1)

CONTRACTS AND PARTICULAR TERMS EXCLUDED FROM THE SCOPE OF THESE REGULATIONS

These Regulations do not apply to–

   (a) any contract relating to employment;
   (b) any contract relating to succession rights;
   (c) any contract relating to rights under family law;
   (d) any contract relating to the incorporation and organisation of companies or partnerships; and
   (e) any term incorporated in order to comply with or which reflects–
      (i) statutory or regulatory provisions of the United Kingdom; or
      (ii) the provisions or principles of international conventions to which the member States or the Community are party.

SCHEDULE 2 Regulation 4(3)

ASSESSMENT OF GOOD FAITH

In making an assessment of good faith, regard shall be had in particular to–

   (a) the strength of the bargaining positions of the parties;
   (b) whether the consumer had an inducement to agree to the term;
   (c) whether the goods or services were sold or supplied to the special order of the consumer, and
   (d) the extent to which the seller or supplier has dealt fairly and equitably with the consumer.

SCHEDULE 3 Regulation 4(4)

INDICATIVE AND ILLUSTRATIVE LIST OF TERMS WHICH MAY BE REGARDED AS UNFAIR

1. Terms which have the object or effect of–

   (a) excluding or limiting the legal liability of a seller or supplier in the event of the death of a consumer or personal injury to the latter resulting from an act or omission of that seller or supplier;

(b) inappropriately excluding or limiting the legal rights of the consumer vis-à-vis the seller or supplier or another party in the event of total or partial non-performance or inadequate performance by the seller or supplier of any of the contractual obligations, including the option of offsetting a debt owed to the seller or supplier against any claim which the consumer may have against him;

(c) making an agreement binding on the consumer whereas provision of services by the seller or supplier is subject to a condition whose realisation depends on his own will alone;

(d) permitting the seller or supplier to retain sums paid by the consumer where the latter decides not to conclude or perform the contract, without providing for the consumer to receive compensation of an equivalent amount from the seller or supplier where the latter is the party cancelling the contract;

(e) requiring any consumer who fails to fulfil his obligation to pay a disproportionately high sum in compensation;

(f) authorising the seller or supplier to dissolve the contract on a discretionary basis where the same facility is not granted to the consumer, or permitting the seller or supplier to retain the sums paid for services not yet supplied by him where it is the seller or supplier himself who dissolves the contract;

(g) enabling the seller or supplier to terminate a contract of indeterminate duration without reasonable notice except where there are serious grounds for doing so;

(h) automatically extending a contract of fixed duration where the consumer does not indicate otherwise, when the deadline fixed for the consumer to express this desire not to extend the contract is unreasonably early;

(i) irrevocably binding the consumer to terms with which he had no real opportunity of becoming acquainted before the con- clusion of the contract;

(j) enabling the seller or supplier to alter the terms of the contract unilaterally without a valid reason which is specified in the contract;

(k) enabling the seller or supplier to alter unilaterally without a valid reason any characteristics of the product or service to be provided;

(l) providing for the price of goods to be determined at the time of delivery or allowing a seller of goods or supplier of services to increase their price without in both cases giving the consumer the corresponding right to cancel the contract if the final price is too high in relation to the price agreed when the contract was concluded;

(m) giving the seller or supplier the right to determine whether the goods or services supplied are in conformity with the contract, or giving him the exclusive right to interpret any term of the contract;

(n) limiting the seller's or supplier's obligation to respect commitments undertaken by his agents or making his commitments subject to compliance with a particular formality;

(o) obliging the consumer to fulfil all his obligations where the seller or supplier does not perform his;

(p) giving the seller or supplier the possibility of transferring his rights and obligations under the contract, where this may serve to reduce the guarantees for the consumer, without the latter's agreement;

(q) excluding or hindering the consumer's right to take legal action or exercise any other legal remedy, particularly by requiring the consumer to take disputes exclusively to arbitration not covered by legal provisions, unduly restricting the evidence available to him or imposing on him a burden of proof which, according to the applicable law, should lie with another party to the contract.

2. Scope of subparagraphs l(g), (j) and (l)

(a) Subparagraph l(g) is without hindrance to terms by which a supplier of financial services reserves the right to terminate unilaterally a contract of indeterminate duration without notice where there is a valid reason, provided that the supplier is required to inform the other contracting party or parties thereof immediately.

(b) Subparagraph l(j) is without hindrance to terms under which a supplier of financial services reserves the right to alter the rate of interest payable by the consumer or due to the latter, or the amount of other charges for financial services without notice where there is a valid reason, provided that the supplier is required to inform the other contracting party or parties thereof at the earliest opportunity and that the latter are free to dissolve the contract immediately.

Subparagraph l(j) is also without hindrance to terms under which a seller or supplier reserves the right to alter unilaterally the conditions of a contract of indeterminate duration, provided that he is required to inform the consumer with reasonable notice and that the consumer is free to dissolve the contract.

(d) Subparagraphs l(g), (j) and (l) do not apply to:
  – transactions in transferable securities, financial instruments and other products or services where the price is linked to fluctuations in a stock exchange quotation or index or a

financial market rate that the seller or supplier does not control;

– contracts for the purchase or sale of foreign currency, traveller's cheques or international money orders denominated in foreign currency;

(d) Subparagraph 1(l) is without hindrance to price indexation clauses, where lawful, provided that the method by which prices vary is explicitly described.

EXPLANATORY NOTE

(This note is not part of the Regulations)

These Regulations implement Council Directive 93/13/EEC on unfair terms in consumer contracts (OJ No. L95, 21.4.93, p. 29).

The Regulations apply, with certain exceptions, to any term which has not been individually negotiated in contracts concluded between a consumer and a seller or supplier (regulation 3). Schedule 1 contains a list of contracts and particular terms which are excluded from the scope of the Regulations. In addition, those terms which define the main subject matter of the contract or concern the adequacy of the price or remuneration as against the goods or services supplied are not to be subject to assessment for fairness, provided that they are in plain, intelligible language (regulation 3(2)).

The Regulations provide that an unfair term is one which contrary to the requirement of good faith causes a significant imbalance in the parties' rights and obligations under the contract to the detriment of the consumer (regulation 4(1)). Schedule 2 contains a list of some of the matters which shall be considered when making an assessment of good faith. Unfair terms are not binding on the consumer (regulation 5).

The Regulations provide that the Director General of Fair Trading shall consider any complaint made to him about the fairness of any contract term drawn up for general use. He may, if he considers it appropriate to do so, seek an injunction to prevent the continued use of that term or a term having like effect in contracts drawn up for general use by a party to the proceedings (regulation 8). In addition, the Director General is given the power to arrange for the dissemination of information and advice concerning the operation of the Regulations (regulation 8(7)).

A compliance cost assessment is available, copies of which have been placed in the libraries of both Houses of Parliament. Copies of the assessment are also available from the Consumer Affairs Division of the Department of Trade and Industry, Room 414, 10–18 Victoria Street, London SW1H 0NN.

Commencement: 1 July 1995.

APPENDIX 6
COUNCIL DIRECTIVE 93/13/EEC

of 5 April 1993
on Unfair Terms in Consumer Contracts

THE COUNCIL OF THE EUROPEAN COMMUNITIES

Having regard to the Treaty establishing the European Economic Community, and in particular Article 100 A thereof,

Having regard to the proposal from the Commission,

In cooperation with the European Parliament,

Having regard to the opinion of the Economic and Social committee,

Whereas it is necessary to adopt measures with the aim of progressively establishing the internal market before 31 December 1992; whereas the internal market comprises an area without internal frontiers in which goods, persons, services and capital move freely;

Whereas the laws of Member States relating to the terms of contract between the seller of goods or supplier of services, on the one hand, and the consumer of them, on the other hand, show many disparities, with the result that the national markets for the sale of goods and services to consumers differ from each other and that distortions of competition may arise amongst the sellers and suppliers, notably when they sell and supply in other Member States;

Whereas, in particular, the laws of Member States relating to unfair terms in consumer contracts show marked divergences;

Whereas it is the responsibility of the Member States to ensure that contracts concluded with consumers do not contain unfair terms;

Whereas, generally speaking, consumers do not know the rules of law which, in Member States other than their own, govern contracts for the sale of goods or services; whereas this lack of awareness may deter them from direct transactions for the purchase of goods or services in another Member State;

Whereas, in order to facilitate the establishment of the internal market and to safeguard the citizen in his role as consumer when acquiring goods and services under contracts which are governed by the laws of Member States other than his own, it is essential to remove unfair terms from those contracts;

Whereas sellers of goods and suppliers of services will thereby be helped in their task of selling goods and supplying services, both at home and throughout the internal market; whereas competition will thus be stimulated, so contributing to increased choice for Community citizens as consumers;

Whereas the two Community programmes for a consumer protection and information policy underlined the importance of safeguarding consumers in the matter of unfair terms of contract; whereas this protection ought to be provided by laws and regulations which are either harmonized at Community level or adopted directly at that level;

Whereas in accordance with the principle laid down under the heading 'Protection of the economic interests of the consumers', as stated in those programmes: 'acquirers of goods and services should be protected against the abuse of power by the seller or supplier, in particular against one-sided standard contracts and the unfair exclusion of essential rights in contracts';

Whereas more effective protection of the consumer can be achieved by adopting uniform rules of law in the matter of unfair terms; whereas those rules should apply to all contracts concluded between sellers or suppliers and consumers; whereas as a result *inter alia* contracts relating to employment, contracts relating to succession rights, contracts relating to rights under family law and contracts relating to the incorporation and organization of companies or partnership agreements must be excluded from this Directive;

Whereas the consumer must receive equal protection under contracts concluded by word of mouth and written contracts regardless, in the latter case, of whether the terms of the contract are contained in one or more documents;

Whereas, however, as they now stand, national laws allow only partial harmonization to be envisaged; whereas, in particular, only contractual terms which have not been individually negotiated are covered by this Directive; whereas Member States should have the option, with due regard for the Treaty, to afford consumers a higher level of protection through national provisions that are more stringent than those of this Directive;

Whereas the statutory or regulatory provisions of the Member States which directly or indirectly determine the terms of consumer contracts are presumed not to contain unfair terms; whereas, therefore, it does not appear to be necessary to subject the terms which reflect mandatory statutory or regulatory provisions and the principles or provisions of international conventions to which the Member States or the Community are party; whereas in that respect wording 'mandatory statutory or regulatory provisions' in Article 1 (2) also covers rules which, according to the law, shall apply between the contracting parties provided that no other arrangements have been established;

Whereas Member States must however ensure that unfair terms are not included, particularly because this Directive also applies to trades, business or professions of a public nature;

Whereas it is necessary to fix in a general way the criteria for assessing the unfair character of contract terms;

Whereas the assessment, according to the general criteria chosen, of the unfair character of terms, in particular in sale or supply activities of a public nature providing collective services which take account of solidarity among users, must be supplemented by a means of making an overall evaluation of the different interests involved; whereas this constitutes the requirement of good faith; whereas, in making an assessment of good faith, particular regard shall be had to the strength of the bargaining positions of the parties, whether the consumer had an inducement to agree to the term and whether the goods or services were sold or supplied to the special order of the consumer; whereas the requirement of good faith may be satisfied by the seller or supplier where he deals fairly and equitably with the other party whose legitimate interests he has to take into account;

Whereas, for the purposes of this Directive, the annexed list of terms can be of indicative value only and, because of the cause of the minimal character of the Directive, the scope of these terms may be the subject of amplification or more restrictive editing by the Member States in their national laws;

Whereas the nature of goods or services should have an influence on assessing the unfairness of contractual terms;

Whereas, for the purposes of this Directive, assessment of unfair character shall not be made of terms which describe the main subject matter of the contract nor the quality/price ratio of the goods or services supplied; whereas the main subject matter of the contract and the price/quality ratio may nevertheless be taken into account in assessing the fairness of other terms; whereas it follows, *inter alia*, that in insurance contracts, the terms which clearly define or circumscribe the insured risk and the insurer's liability shall not be subject to such assessment since these restrictions are taken into account in calculating the premium paid by the consumer;

Whereas contracts should be drafted in plain, intelligible language, the consumer should actually be given an opportunity to examine all the terms and, if in doubt, the interpretation most favourable to the consumer should prevail;

Whereas Member States should ensure that unfair terms are not used in contracts concluded with consumers by a seller or supplier and that if, nevertheless, such terms are so used, they will not bind the consumer, and the contract will continue to bind the parties upon those terms if it is capable of continuing in existence without the unfair provisions;

Whereas there is a risk that, in certain cases, the consumer may be deprived of protection under this Directive by designating the law of a non-Member country as the law applicable to the contract; whereas provisions should therefore be included in this Directive designed to avert this risk;

Whereas persons or organizations, if regarded under the law of a Member State as having a legitimate interest in the matter, must have facilities for initiating proceedings concerning terms of contract drawn up for general use in contracts concluded with consumers, and in particular unfair terms, either before a court or before an administrative authority competent to decide upon complaints or to initiate appropriate legal proceedings; whereas this possibility does not, however, entail prior verification of the general conditions obtaining in individual economic sectors;

Whereas the courts or administrative authorities of the Member States must have at their disposal adequate and effective means of preventing the continued application of unfair terms in consumer contracts,

HAS ADOPTED THIS DIRECTIVE:

Article 1

1. The purpose of this Directive is to approximate the laws, regulations and administrative provisions of the Member States relating to unfair terms in contracts concluded between a seller or supplier and a consumer.

2. The contractual terms which reflect mandatory statutory or regulatory provisions and the provisions or principles of international conventions to which the Member States or the Community are party, particularly in the transport area, shall not be subject to the provisions of this Directive.

Article 2

For the purposes of this Directive:

    (a) 'unfair terms' means the contractual term defined in Article 3;
    (b) 'consumer' means any natural person who, in contracts covered by this Directive, is acting for purposes which are outside his trade, business or profession;
    (c) 'seller or supplier' means any natural or legal person who, in contracts covered by this Directive, is acting for purposes relating to his trade, business or profession, whether publicly owned or privately owned.

Article 3

1. A contractual term which has not been individually negotiated shall be regarded as unfair if, contrary to the requirement of good faith, it causes a significant imbalance in the parties' rights and

obligations arising under the contract, to the detriment of the consumer.

2. A term shall always be regarded as not individually negotiated where it has been drafted in advance and the consumer has therefore not been able to influence the substance of the term, particularly in the context of a pre-formulated standard contract.

The fact that certain aspects of a term or one specific term have been individually negotiated shall not exclude the application of this Article to the rest of a contract if an overall assessment of the contract indicates that it is nevertheless a pre-formulated standard contract.

Where any seller or supplier claims that a standard term has been individually negotiated, the burden of proof in this respect shall be incumbent on him.

3. The Annex shall contain an indicative and non-exhaustive list of the terms which may be regarded as unfair.

Article 4

1. Without prejudice to Article 7, the unfairness of a contractual term shall be assessed, taking into account the nature of the goods or services for which the contract was concluded and by referring, at the time of conclusion of the contract, to all the circumstances attending the conclusion of the contract and to all the other terms of the contract or of another contract on which it is dependent.

2. Assessment of the unfair nature of the terms shall relate neither to the definition of the main subject matter of the contract nor to the adequacy of the price and remuneration, on the one hand, as against the services or goods supplied in exchange, on the other, in so far as these terms are in plain intelligible language.

Article 5

In the case of contracts where all or certain terms offered to the consumer are in writing, these terms must always be drafted in plain, intelligible language. Where there is doubt about the meaning of a term, the interpretation most favourable to the consumer shall prevail. This rule on interpretation shall not apply in the context of the procedures laid down in Article 7 (2).

Article 6

1. Member States shall lay down that unfair terms used in a contract concluded with a consumer by a seller or supplier shall, as provided for under their national law, not be binding on the consumer and that the contract shall continue to bind the parties upon those terms if it is capable of continuing in existence without the unfair terms.

2. Member States shall take the necessary measures to ensure that the consumer does not lose the protection granted by this Directive by virtue of the choice of the law of a non-Member country as the law applicable to the contract if the latter has a close connection with the territory of the Member States.

Article 7

1. Member States shall ensure that, in the interests of consumers and of competitors, adequate and effective means exist to prevent the continued use of unfair terms in contracts concluded with consumers by sellers or suppliers.

2. The means referred to in paragraph 1 shall include provisions whereby persons or organizations, having a legitimate interest under national law in protecting consumers, may take action according to the national law concerned before the courts or before competent administrative bodies for a decision as to whether contractual terms drawn up for general use are unfair, so that they can apply appropriate and effective means to prevent the continued use of such terms.

3. With due regard for national laws, the legal remedies referred to in paragraph 2 may be directed separately or jointly against a number of sellers or suppliers from the same economic sector or their associations which use or recommend the use of the same general contractual terms or similar terms.

Article 8

Member States may adopt or retain the most stringent provisions compatible with the Treaty in the area covered by this Directive, to ensure a maximum degree of protection for the consumer.

Article 9

The Commission shall present a report to the European Parliament and to the Council concerning the application of this Directive five years at the latest after the date in Article 10 (1).

Article 10

1. Member States shall bring into force the laws, regulations and administrative provisions necessary to comply with this Directive no later than 31 December 1994. They shall forthwith inform the Commission thereof.

These provisions shall be applicable to all contracts concluded after 31 December 1994.

2. When Member States adopt these measures, they shall contain a reference to this Directive or shall be accompanied by such reference on the occasion of their official publication. The methods of making such a reference shall be laid down by the Member States.

3. Member States shall communicate the main provisions of national law which they adopt in the field covered by this Directive to the Commission.

Article 11

This Directive is addressed to the Member States.

Done at Luxembourg, 5 April 1993.

For the Council
The President
N. HELVEG PETERSEN

ANNEX
TERMS REFERRED TO IN ARTICLE 3 (3)

1. Terms which have the object or effect of:

(a) excluding or limiting the legal liability of a seller or supplier in the event of the death of a consumer or personal injury to the latter resulting from an act or omission of that seller or supplier;

(b) inappropriately excluding or limiting the legal rights of the consumer *vis-à-vis* the seller or supplier or another party in the event of total or partial non-performance or inadequate performance by the seller or supplier of any of the contractual obligations, including the option of offsetting a debt owed to the seller or supplier against any claim which the consumer may have against him;

(c) making an agreement binding on the consumer whereas provision of services by the seller or supplier is subject to a condition whose realization depends on his own will alone;

(d) permitting the seller or supplier to retain sums paid by the consumer where the latter decides not to conclude or perform the contract, without providing for the consumer to receive compensation of an equivalent amount from the seller or supplier where the latter is the party cancelling the contract;

(e) requiring any consumer who fails to fulfil his obligation to pay a disproportionately high sum in compensation;

(f) authorizing the seller or supplier to dissolve the contract on a discretionary basis where the same facility is not granted to the consumer, or permitting the seller or supplier to retain the sums paid for services not yet supplied by him where it is the seller or supplier himself who dissolves the contract;

(g) enabling the seller or supplier to terminate a contract of indeterminate duration without reasonable notice except where there are serious grounds for doing so;

(h) automatically extending a contract of fixed duration where the consumer does not indicate otherwise, when the deadline fixed for the consumer to express this desire not to extend the contract is unreasonably early;

(i) irrevocably binding the consumer to terms with which he had no real opportunity of becoming acquainted before the conclusion of the contract;

(j) enabling the seller or supplier to alter the terms of the contract unilaterally without a valid reason which is specified in the contract;

(k) enabling the seller or supplier to alter unilaterally without a valid reason any characteristics of the product or service to be provided;

(l) providing for the price of goods to be determined at the time of delivery or allowing a seller of goods or supplier of services to increase their price without in both cases giving the consumer the corresponding right to cancel the contract if the final price is too high in relation to the price agreed when the contract was concluded;

(m) giving the seller or supplier the right to determine whether the goods or services supplied are in conformity with the contract, or giving him the exclusive right to interpret any term of the contract;

(n) limiting the seller's or supplier's obligation to respect commitments undertaken by his agents or making his commitments subject to compliance with a particular formality;

(o) obliging the consumer to fulfil all his obligations where the seller or supplier does not perform his;

(p) giving the seller or supplier the possibiliity of transferring his rights and obligations under the contract, where this may serve to reduce the guarantees for the consumer, without the latter's agreement;

(q) excluding or hindering the consumer's right to take legal action or exercise any other legal remedy, particularly by requiring the consumer to take disputes exclusively to arbitration not covered by legal provisions, unduly restricting the evidence available to him or imposing on him a burden of proof which, according to the applicable law, should lie with another party to the contract.

2. Scope of subparagraphs (g), (j) and (l)

(a) Subparagraph (g) is without hindrance to terms by which a supplier of financial services reserves the right to terminate unilaterally a contract of indeterminate duration without notice where there is a valid reason, provided that the supplier is required to inform the other contracting party or parties thereof immediately.

(b) Subparagraph (j) is without hindrance to terms under which a supplier of financial services reserves the right to alter the rate of interest payable by the consumer or due to the latter, or the amount of other charges for financial services without notice where there is a valid reason, provided that the supplier is rquired to inform the other contracting party or parties thereof at the earliest opportunity and that the latter are free to dissolve the contract immediately.

Subparagraph (j) is also without hindrance to terms under which a seller or supplier reserves the right to alter unilaterally the conditions of a contract of indeterminate duration, provided that he is required to inform the consumer with reasonable notice and that the consumer is free to dissolve the contract.

(c) Subparagraphs (g), (j) and (l) do not apply to:
   - transactions in transferable securities, financial instruments and other products or services where the price is linked to fluctuations in a stock exchange quotation or index or a financial market rate that the seller or supplier does not control;
   - contracts for the purchase or sale of foreign currency, traveller's cheques or international money orders denominated in foreign currency;

(d) Subparagraph (l) is without hindrance to price-indexation clauses, where lawful, provided that the method by which prices vary is explicitly described.

APPENDIX 7
SET-OFF CLAUSES FROM SELECTED STANDARD FORMS OF CONTRACT

1. JCT Standard Form of Building Contract 1998

24   Damages for non-completion

24.1 If the Contractor fails to complete the Works by the Completion Date then the Architect shall issue a certificate to that effect. In the event of a new Completion Date being fixed after the issue of such a certificate such fixing shall cancel that certificate and the Architect shall issue such further certificate under clause 24.1 as may be necessary.

Certificate of Architect

24.2 .1. Provided:
   – the Architect has issued a certificate under clause 24.1; and
   – the Employer has informed the Contractor in writing before the date of the Final Certificate that he may require payment of, or may withhold or deduct, liquidated and ascertained damages,
   the Employer may, not later than 5 days before the final date for payment of the debt due under the Final Certificate:

Payment or allowance of liquidated damages

either

   .1 .1 require in writing the Contractor to pay to the Employer liquidated and ascertained damages at the rate stated in the Appendix (or at such lesser rate as may be specified in writing by the Employer) for the period between the Completion Date and the date of Practical Completion and the Employer may recover the same as a debt;

or

   .1 .2 give a notice pursuant to clause 30.1.1.4 or clause 30.8.3 to the Contractor that he will deduct from monies due to the Contractor liquidated and ascertained damages at the rate stated in the Appendix (or at such lesser rate as may be specified in the notice) for the period between the Completion Date and the date of Practical Completion.

24.2 .2 If, under clause 25.3.3, the Architect fixes a later Completion Date or a later Completion Date is stated in a confirmed acceptance of a 13A Quotation, the Employer shall pay or repay to the Contractor any amounts recovered, allowed or paid under clause 24.2.1 for the period up to such later Completion Date.

24.2 .3 Notwithstanding the issue of any further certificate of the Architect under clause 24.1 any requirement of the Employer which has been previously stated in writing in accordance with clause 24.2.1 shall remain effective unless withdrawn by the Employer.

30 Certificates and payments

Interim
Certificates and
valuations – final
date for payment
– interest

30.1 .1 .1 The Architect shall from time to time as provided in clause 30 issue Interim Certificates stating the amount due to the Contractor from the Employer specifying to what the amount relates and the basis on which that amount was calculated; and the final date for payment pursuant to an Interim Certificate shall be 14 days from the date of issue of each Interim Certificate.

If the Employer fails properly to pay the amount, or any part thereof, due to the Contractor under the Conditions by the final date for its payment the Employer shall pay to the Contractor in addition to the amount not properly paid simple interest thereon for the period until such payment is made. Payment of such simple interest shall be treated as a debt due to the Contractor by the Employer. The rate of interest payable shall be five per cent (5%) over the Base Rate of the Bank of England which is current at the date the payment by the Employer became overdue. Any payment of simple interest under this clause 30.1.1.1 shall not in any circumstances be construed as a waiver by the Contractor of his right to proper payment of the principal amount due from the Employer to the Contractor in accordance with, and within the time stated in, the Conditions or of the rights of the Contractor in regard to suspension of the performance of his obligations under this Contract to the Employer pursuant to clause 30.1.4 or to determination of his employment pursuant to the default referred to in clause 28.2.1.1.

.1 .2 Notwithstanding the fiduciary interest of the Employer in the Retention as stated in clause 30.5.1 the Employer is entitled to exercise any right under this Contract of withholding and/or deduction from monies due or to become due to the Contractor against any amount so due under an Interim Certificate whether or not any Retention is included in that Interim Certificate by the operation of clause 30.4. Such withholding and/or deduction is subject to the restriction in clause 35.13.5.3.2

.1 .3 Not later than 5 days after the date of issue of an Interim Certificate the Employer shall give a written notice to the Contractor which shall, in respect of the amount stated as due in that Interim Certificate, specify the amount of the payment proposed to be made, to what the amount of the payment relates and the basis on which that amount is calculated.

.1 .4 Not later than 5 days before the final date for payment of the amount due pursuant to clause 30.1.1.1 the Employer may give a written notice to the Contractor which shall specify any amount proposed to be withheld and/or deducted from that due amount, the ground or grounds for such withholding and/or deduction and the amount of withholding and/or deduction attributable to each ground.

.1 .5 Where the Employer does not give any written notice pursuant to clause 30.1.1.3 and/or to clause 30.1.1.4 the Employer shall pay the Contractor the amount due pursuant to clause 30.1.1.1.

30.1 .1 .6 Where it is stated in the Appendix that clause 30.1.1.6 applies, the advance payment identified in the Appendix shall be paid to the Contractor on the date stated in the Appendix and such advance payment shall be reimbursed to the Employer by the Contractor on the terms stated in the Appendix. Provided that where the Appendix states that an advance payment bond is required such payment shall only be made if the Contractor has provided to the Employer such bond from a surety approved by the Employer on the terms agreed between the British Bankers' Association and the JCT and annexed to the Appendix unless pursuant to the Seventh Recital a bond on other terms is required by the Employer. **Advance payment**

30.1 .2 .1 Interim valuations shall be made by the Quantity Surveyor whenever the Architect considers them to be necessary for the purpose of ascertaining the amount to be stated as due in an Interim Certificate. **Interim valuations**

.2 .2 Without prejudice to the obligation of the Architect to issue Interim Certificates as stated in clause 30.1.1.1, the Contractor, not later than 7 days before the date of an Interim Certificate, may submit to the Quantity Surveyor an application which sets out what the Contractor considers to be the amount of the gross valuation pursuant to clause 30.2. The Contractor shall include with his application any application made to the Contractor by a Nominated Sub-Contractor which sets out what the Nominated Sub-Contractor considers to be the amount of the gross valuation pursuant to clause 4.17 of Conditions NSC/C. If the Contractor submits such an application the Quantity Surveyor shall make an **Application by Contractor – amount of gross valuation** interim valuation. To the extent that the Quantity Surveyor disagrees with the gross valuation in the Contractor's application and/or in a Nominated Sub-contractor's application the Quantity Surveyor at the same time as making the valuation shall submit to the Contractor a statement, which shall be in similar detail to that given in the application, which identifies such disagreement.

Issue of Interim Certificates

30.1 .3 Interim Certificates shall be issued at the Period of Interim Certificates specified in the Appendix up to and including the end of the period during which the certificate of Practical Completion is issued. Thereafter Interim Certificates shall be issued as and when further amounts are ascertained as payable to the Contractor from the Employer and after the expiration of the Defects Liability Period named in the Appendix or upon the issue of the Certificate of Completion of Making Good Defects (whichever is the later) provided always that the Architect shall not be required to issue an Interim Certificate within one calendar month of having issued a previous Interim Certificate.

Right of suspension of obligations by Contractor

30.1 .4 Without prejudice to any other rights and remedies which the Contractor may possess, if the Employer shall, subject to any notice issued pursuant to clause 30.1.1.4, fail to pay the Contractor in full (including any VAT due pursuant to the VAT Agreement) by the final date for payment as required by the Conditions and such failure shall continue for 7 days after the Contractor has given to the Employer, with a copy to the Architect, written notice of his intention to suspend the performance of his obligations under this Contract to the Employer and the ground or grounds on which it is intended to suspend performance then the Contractor may suspend such performance of his obligations under this Contract to the Employer until payment in full occurs. Such suspension shall not be treated as a suspension to which clause 27.2.1.1 refers or a failure to proceed regularly and diligently with the Works to which clause 27.2.1.2 refers.

Interim Certificate – final adjustment or ascertainment of nominated subcontract sums

30.7 So soon as is practicable but not less than 28 days before the date of issue of the Final Certificate referred to in clause 30.8 and notwithstanding that a period of one month may not have elapsed since the issue of the previous Interim Certificate, the Architect shall issue an Interim Certificate the gross valuation for which shall include the amounts of the sub-contract sums for all Nominated Sub-Contracts as finally adjusted or ascertained under all relevant provisions of Conditions NSC/C.

Issue of Final Certificate

30.8 .1 The Architect shall issue the Final Certificate (and inform each Nominated Sub-Contractor of the date of its issue) not later than 2 months after whichever of the following occurs last;

the end of the Defects Liability Period;

the date of issue of the Certificate of Completion of Making Good Defects under clause 17.4;

the date on which the Architect sent a copy to the Contractor of any ascertainment to which clause 30.6.1.2.1 refers and of the statement prepared in compliance with clause 30.6.1.2.2.

The Final Certificate shall state:

.1 .1 the sum of the amounts already stated as due in Interim Certificates plus the amount of any advance payment paid pursuant to clause 30.1.1.6, and

.1 .2 the Contract Sum adjusted as necessary in accordance with clause 30.6.2, and

.1 .3 to what the amount relates and the basis on which the statement in the Final Certificate has been calculated

and the difference (if any) between the two sums shall (without prejudice to the rights of the Contractor in respect of any Interim Certificates which have subject to any notice issued pursuant to clause 30.1.1.4 not been paid in full by the Employer by the final date for payment of such Certificate) be expressed in the said Certificate as a balance due to the Contractor from the Employer or to the Employer from the Contractor as the case may be.

30.8 .2 Not later than 5 days after the date of issue of the Final Certificate the Employer shall give a written notice to the Contractor which shall, in respect of any balance stated as due to the Contractor from the Employer in the Final Certificate, specify the amount of the payment proposed to be made, to what the amount of the payment relates and the basis on which that amount is calculated.

30.8 .3 The final date for payment of the said balance payable by the Employer to the Contractor or by the Contractor to the Employer as the case may be shall be 28 days from the date of issue of the said Certificate. Not later than 5 days before the final date for payment of the balance the Employer may give a written notice to the Contractor which shall specify any amount proposed to be withheld and/or deducted from any balance due to the Contractor, the ground or grounds for such withholding and/or deduction and the amount of withholding and/or deduction attributable to each ground.

30.8 .4 Where the Employer does not give a written notice pursuant to clause 30.8.2 and/or clause 30.8.3 the Employer shall pay the Contractor the balance stated as due to the Contractor in the Final Certificate.

30.8 .5 If the Employer or the Contractor fails properly to pay the said balance, or any part thereof, by the final date for its payment the Employer or the Contractor as the case may be shall pay to the other, in addition to the balance not properly paid, simple interest thereon for the period until such payment is made. The rate of interest payable shall be five per cent (5%) over the Base Rate of the Bank of England which is current at the date the payment by the Employer or by the Contractor as the case may be became overdue. Any payment of simple interest under this clause 30.8 shall not in any circumstances be construed as a waiver by the Contractor or by the Employer as the case may be of his right to proper payment of the aforesaid balance due from the Employer to the Contractor or from the Contractor to the Employer in accordance with this clause 30.8.

30.8 .6 Liability for payment of the balance pursuant to clause 30.8.3 and of any interest pursuant to clause 30.8.5 shall be treated as a debt due to the Contractor by the Employer or to the Employer by the Contractor as the case may be.

Effect of Final Certificate

30.9 .1 Except as provided in clauses 30.9.2 and 30.9.3 (and save in respect of fraud), the Final Certificate shall have effect in any proceedings under or arising out of or in connection with this Contract (whether by adjudication under article 5 or by arbitration under article 7A or by legal proceedings under article 7B) as

.1 .1 conclusive evidence that where and to the extent that any of the particular qualities of any materials or goods or any particular standard of an item of workmanship was described expressly in the Contract Drawings or the Contract Bills, or in any of the Numbered Documents, or in any instruction issued by the Architect under the Conditions, or in any drawings or documents issued by the Architect under clause 5.3.1.1 or 5.4 or 7, to be for the approval of the Architect, the particular quality or standard was to the reasonable satisfaction of the Architect, but such Certificate shall not be conclusive evidence that such or any other materials or goods or workmanship comply or complies with any other requirement or term of this Contract, and

.1 .2 conclusive evidence that any necessary effect has been given to all the terms of this Contract which require that an amount is to be added to or deducted from the Contract Sum or an adjustment is to be made of the Contract Sum save where there has been any accidental inclusion or exclusion of any work, materials, goods or figure in any computation or any arithmetical error in any computation, in which event the Final Certificate shall have effect as conclusive evidence as to all other computations, and

.1 .3 conclusive evidence that all and only such extensions of time, if any, as are due under clause 25 have been given, and

.1 .4 conclusive evidence that the reimbursement of direct loss and/or expense, if any, to the Contractor pursuant to clause 26.1 is in final settlement of all and any claims which the Contractor has or may have arising out of the occurrence of any of the matters referred to in clause 26.2 whether such claim be for breach of contract, duty of care, statutory duty or otherwise.

30.9 .2 If any adjudication, arbitration or other proceedings have been commenced by either Party before the Final Certificate has been issued the Final Certificate shall have effect as conclusive evidence as provided in clause 30.9.1 after either

.2 .1 such proceedings have been concluded, whereupon the Final Certificate shall be subject to the terms of any decision, award or judgment in or settlement of such proceedings, or

.2 .2 a period of 12 months after the issue of the Final Certificate during which neither Party has taken any further step in such proceedings, whereupon the Final Certificate shall be subject to any terms agreed in partial settlement,

whichever shall be the earlier.

2. JCT Form of Building Contract for Works of Simple Content 1998 Edition – IFC 98

Certificate of non-completion

2.6 If the Contractor fails to complete the Works by the Date for Completion or within any extended time fixed under clause **2.3** then the Architect/the Contract Administrator shall issue a certificate to that effect.

In the event of an extension of time being made after the issue of such a certificate such making shall cancel that certificate and the Architect/ the Contract Administrator shall issue such further certificate under this clause as may be necessary.

Liquidated damages for non-completion

2.7 Provided:
- the Architect/the Contract Administrator has issued a certificate under clause **2.6**; and
- the Employer has informed the Contractor in writing before the date of the final certificate that he may require payment of, or may withhold or deduct, liquidated and ascertained damages.

the Employer may not later than 5 days before the final date for payment of the debt due under the final certificate

either

2.7.1 require in writing the Contractor to pay to the Employer liquidated and ascertained damages at the rate stated in the Appendix for the period during which the Works shall remain or have remained incomplete and may recover the same as a debt

or

2.7.2 give a notice pursuant to clause **4.2.3(b)** or clause **4.6.1.3** that he will deduct liquidated damages at the rate stated in the Appendix for the period during which the Works shall remain or have remained incomplete

Notwithstanding the issue of any further certificate of the Architect/the Contract Administrator under clause **2.6** any written requirement or notice given to the Contractor in accordance with this clause shall remain effective unless withdrawn by the Employer.

Interim payments

4.2 (a) Subject to any agreement between the Parties as to stage payments, the Architect/the Contract Administrator shall, at intervals of one month, unless a different interval is stated in the Appendix, calculated from the

Date of Possession stated in the Appendix, certify the amount of interim payments to be made by the Employer to the Contractor specifying to what the amount relates and the basis on which that amount was calculated and the final date for payment pursuant to a certificate shall be 14 days from the date of issue of each certificate.

4.2.3 (a) Not later than 5 days after the date of issue of a certificate of interim payment, the Employer shall give a written notice to the Contractor which shall, in respect of the amount stated as due in that certificate of interim payment, specify the amount of the payment proposed to be made, to what the amount of the payment related and the basis on which the amount is calculated.

(b) Not later than 5 days before the final date for payment of the amount due pursuant to clauses **4.2.1** and **4.2.2** the Employer may give a written notice to the Contractor which shall specify any amount proposed to be withheld and/or deducted from that due amount, the ground or grounds for such withholding and/or deduction and the amount of the withholding and/or deduction attributable to each ground.

(c) Where the Employer does not give a written notice pursuant to clause **4.2.3(a)** and/or to clause **4.2.3(b)** the Employer shall pay the amount due pursuant to clauses **4.2.1** and **4.2.2**.

Interim payment on Practical Completion

4.3 (a) The Architect/The Contract Administrator shall, within 14 days after the date of Practical Completion, certify payment to be made by the Employer to the Contractor of $97\frac{1}{2}\%$ of the total value referred to in clause **4.2.1(a)** and 100% of any amounts payable pursuant to clause **4.2.2** together with any deduction under clause **3.9** *(Levels)* or **4.9(a)** *(Tax etc. fluctuations)* or **4.10** *(Fluctuations: named persons)* less the amount of any advance payment made pursuant to clause **4.2(b)** and less any sums previously certified for payment. The final date for payment of the amount pursuant to the certificate shall be 14 days from the date of issue of the certificate.

(b) Not later than 5 days after the date of issue of the certificate the Employer shall give a written notice to the Contractor which shall, in respect of the amount stated as due in that certificate, specify the amount of the payment proposed to be made, to what the amount of the payment relates and the basis on which that amount is calculated.

(c) Not later than 5 days before the final date for payment of the amount due pursuant to clause **4.3(a)** the Employer may give a written notice to the Contractor which shall specify any amount proposed to be withheld and/or deducted from that due amount, the ground or grounds for such withholding and/or deduction and the amount of the withholding and/or deduction attributable to each ground.

(d) Where the Employer does not give a written notice pursuant to clause **4.3(b)** and/or to clause **4.3(c)** the Employer shall pay the amount due pursuant to clause **4.3(a)**.

(e) If the Employer fails properly to pay the amount, or any part thereof, due to the Contractor under the Conditions by the final date for its payment the Employer shall pay to the Contractor in addition to the amount not properly paid simple interest thereon on the terms set out in clause **4.2(a)**.

Issue of final certificate

4.6.1 .1 The Architect/The Contract Administrator shall, within 28 days of the sending of such computations of the adjusted Contract Sum to the Contractor or of the certificate issued by the Architect/the Contract Administrator under clause **2.10** *(Defects liability)*, whichever is the later, issue a final certificate certifying the amount due to the Contractor or to the Employer as the case may be and stating to what the amount relates and the basis on which the amount due under the final certificate has been calculated. The amount to be certified shall be the Contract Sum adjusted as stated in clause **4.5** less the amount of any advance payment made pursuant to clause **4.2(b)** and less any sums previously certified for payment.

4.6.1 .2 Not later than 5 days after the date of issue of the final certificate the Employer shall give a written notice to the Contractor which shall, in respect of any amount stated as due to the Contractor from the Employer in the final certificate, specify the amount of the payment proposed to be made, to what the amount of the payment relates and the basis on which that amount is calculated.

4.6.1 .3 The final date for payment of the said amount by the Employer to the Contractor or by the Contractor to the Employer as the case may be shall be 28 days from the date of issue of the said certificate. Not later than 5 days before the final date for payment of the said amount the Employer may give a written notice to the Contractor which shall specify any amount proposed to be withheld and/or deducted from any amount due to the Contractor, the ground or grounds for such withholding and/or deduction and the amount of withholding and/or deduction attributable to each ground.

4.6.1 .4 Where the Employer does not give a written notice pursuant to clause **4.6.1.2** and/or to clause **4.6.1.3** the Employer shall pay the Contractor the amount stated as due to the Contractor in the final certificate.

3. JCT Agreement for Minor Building Works

Damages for non-completion

2.3 If the Works are not completed by the completion date inserted in clause 2.1 hereof or by any later completion date fixed under clause 2.2 hereof the Contractor shall pay or allow to the Employer liquidated damages at the rate of

£.................... per ....

between the aforesaid completion date and the date of practical completion.

The Employer may

either

recover the liquidated damages from the Contractor as a debt

or

deduct the liquidated damages from any monies due to the Contractor under this Contract provided that a notice of deduction pursuant to clause 4.4.2 or clause 4.5.1.3 has been given. If the Employer intends to deduct any such damages from the sum stated as due in the final certificate, he shall additionally inform the Contractor, in writing, of that intention not later than the date of issue of the final certificate.

Progress payments and retention

4.2 .1 The Architect/The Contract Administrator shall at intervals of 4 weeks calculated from the date for commencement, certify progress payments as amounts due to the Contractor in respect of the value of the Works properly executed, including any amounts either ascertained or agreed under clauses 3.6 and 3.7 hereof, and the value of any materials and goods which have been reasonably and properly brought upon the site for the purpose of the Works and which are adequately stored and protected against the weather and other casualties, less a retention of 5%/.........% and less the total amounts due to the Contractor in certificates of progress payment previously issued. The certificate shall state to what the progress payment relates and the basis on which the amount of the progress payment was calculated. The final date for payment by the Employer of the amount so certified shall be 14 days from the date of issue of the certificate. The provisions of clause 4.4 shall apply to any certificate issued pursuant to this clause 4.2.1.

.2 If the Employer fails properly to pay the amount, or any part thereof, due to the Contractor by the final date for its payment the Employer shall pay to the Contractor in addition to the amount not properly paid simple interest thereon for the period until such payment is made. Payment of such simple interest shall be treated as a debt due

163

to the Contractor by the Employer. The rate of interest payable shall be five per cent (5%) over the Base Rate of the Bank of England which is current at the date the payment by the Employer became overdue. Any payment of simple interest under this clause 4.2.2 shall not in any circumstances be construed as a waiver by the Contractor of his right to proper payment of the principal amount due from the Employer to the Contractor in accordance with, and within the time stated in, the Conditions or of the rights of the Contractor in regard to suspension of performance of his obligations under this Agreement to the Employer pursuant to clause 4.8 or to determination of his employment pursuant to the default referred to in clause 7.3.1.1.

Penultimate certificate

4.3 The Architect/The Contract Administrator shall within 14 days after the date of practical completion certified under clause 2.4 hereof certify payment as an amount due to the Contractor of $97\frac{1}{2}\%/\ldots\ldots\%$ of the total amount to be paid to the Contractor under this Contract so far as that amount is ascertainable at the date of practical completion including any amounts either ascertained or agreed under clauses 3.6 and 3.7 hereof less the total amounts due to the Contractor in certificates of progress payment previously issued. The penultimate certificate shall state to what the progress payment relates and the basis on which the amount of the certificate was calculated. The final date for payment by the Employer of the amount so certified shall be 14 days from the date of issue of that certificate. If the Employer fails properly to pay the amount, or any part thereof, due to the Contractor by the final date for its payment the provisions of clause 4.2.2 shall apply. The provisions of clause 4.4 shall apply to the certificate issued pursuant to this clause 4.3.

Notices of amounts to be paid and deductions

4.4 .1 Not later than 5 days after the issue of a certificate of payment pursuant to clauses 4.2.1 and 4.3 the Employer shall give a written notice to the Contractor which shall specify the amount of the payment proposed to be made in respect of the amount stated as due in that certificate.

.2 Not later than 5 days before the final date for payment of the amount notified pursuant to clause 4.4.1 the Employer may give a written notice to the Contractor which shall specify any amount proposed to be withheld and/or deducted from that notified amount, the ground or grounds for such withholding and/or deduction and the amount of the withholding and/or deduction attributable to each ground.

.3 Where the Employer does not give a written notice pursuant to clause 4.4.1 and/or to clause 4.4.2 the Employer shall pay the amount stated as due in the certificate.

Final certificate

4.5 .1 .1 The Contractor shall supply within 3 months/.............from the date of practical completion all documentation reasonably required for the computation of the amount to be finally certified by the Architect/the Contract Administrator and the Architect/the Contract Administrator shall within 28 days of receipt of such documentation, provided that the Architect/the Contract Administrator has issued the certificate under clause 2.5 hereof, issue a final certificate certifying the amount remaining due to the Contractor or due to the Employer as the case may be and shall state to what the amount relates and the basis on which that amount was calculated.

.2 Not later than 5 days after the date of issue of the final certificate the Employer shall give a written notice to the Contractor which shall specify the amount of the payment proposed to be made to the Contractor in respect of the amount certified.

.3 The final date for payment of such amount as a debt payable as the case may be by the Employer to the Contractor or by the Contractor to the Employer shall be 14 days from the date of issue of the said certificate. Not later than 5 days before the final date for payment of the amount due to the Contractor the Employer may give a written notice to the Contractor which shall specify any amount proposed to be withheld and/or deducted therefrom, the ground or grounds for such withholding and/or deduction and the amount of the withholding and/or deduction attributable to each ground.

.4 Where the Employer does not give a written notice pursuant to clause 4.5.1.2 and/or the clause 4.5.1.3 the Employer shall pay the Contractor the amount stated as due to the Contractor in the final certificate.

.2 If the Employer or the Contractor fails properly to pay the debt, or any part thereof, by the final date for its payment the Employer or the Contractor as the case may be shall pay to the other in addition to the debt not properly paid simple interest thereon for the period until such payment is made. The rate of interest payable shall be five per cent (5%) over the Base Rate of the Bank of England which is current at the date the payment by the Employer or by the Contractor as the case may be became overdue. Any payment of simple interest under this clause 4.5.2 shall not in any circumstances be construed as a waiver by the Contractor or by the Employer as the case may be of his right to proper payment of the aforesaid debt due from the Employer to the Contractor or from the Contractor to the Employer in accordance with clause 4.5.1.

4. ICE Conditions of Contract 6th Edition

Clauses 60 and 66 as amended

Clause 60 – Certificates and Payment

Monthly statements

(1) Unless otherwise agreed the Contractor shall submit to the Engineer at monthly intervals commencing within one month after the Works Commencement Date a statement (in such form if any as may be prescribed in the Specification) showing

(a) the estimated contract value of the Permanent Works executed up to the end of that month

(b) a list of any goods or materials delivered to the Site for but not yet incorporated in the Permanent Works and their value

(c) a list of any of those goods or materials identified in the Appendix to the Form of Tender which have not yet been delivered to the Site but of which the property has vested in the Employer pursuant to Clause 54 and their value and

(d) the estimated amounts to which the Contractor considers himself entitled in connection with all other matters for which provision is made under the Contract including any Temporary Works or Contractor's Equipment for which separate amounts are included in the Bill of Quantities unless in the opinion of the Contractor such values and amounts together will not justify the issue of an interim certificate.

Amounts payable in respect of Nominated Sub-contracts are to be listed separately.

Monthly payments

(2) Within 25 days of the date of delivery of the Contractor's monthly statement to the Engineer or the Engineer's representative in accordance with sub-clause (1) of this clause the Engineer shall certify and within 28 days of the same date the Employer shall pay to the Contractor (after deducting any previous payments on account)

(a) the amount which in the opinion of the Engineer on the basis of the monthly statement is due to the Contractor on account of sub-clauses (1)(a) and (1)(d) of this Clause less a retention as provided in sub-clause (5) of this Clause and

(b) such amounts (if any) as the Engineer may consider proper (but in no case exceeding the percentage of the value stated in the Appendix to the Form of Tender) in respect of sub-clauses (1)(b) and (1)(c) of this Clause.

The payments become due on certification with the final date for payment being 28 days after the date of delivery of the Contractor's monthly statement.

The amounts certified in respect of Nominated Sub-contracts shall be shown separately in the certificate.

Minimum amount of certificate

(3) Until the whole of the Works has been certified as substantially complete in accordance with Clause 48 the Engineer shall not be bound to issue an interim certificate for a sum less than that stated in the Appendix to the Form of Tender but thereafter he shall be bound to do so and the certification and payment of amounts due to the Contractor shall be in accordance with the time limits contained in this Clause.

Final account

(4) Not later than 3 months after the date of the Defects Correction Certificate the Contractor shall submit to the Engineer a statement of final account and supporting documentation showing in detail the value in accordance with the Contract of the Works executed together with all further sums which the Contractor considers to be due to him under the Contract up to the date of the Defects Correction Certificate.

Within 3 months after receipt of this final account and of all information reasonably required for its verification the Engineer shall issue a certificate stating the amount which in his opinion is finally due under the Contract from the Employer to the Contractor or from the Contractor to the Employer as the case may be up to the date of the Defects Correction Certificate and after giving credit to the Employer for all amounts previously paid by the Employer and for all sums to which the Employer is entitled under the Contract. Such amount shall subject to Clause 47 be paid to or by the Contractor as the case may require. The payment becomes due on certification. The final date for payment is 28 days later.

Retention

(5) The retention to be made pursuant to sub-clause (2)(a) of this Clause shall be the difference between.

   (a) an amount calculated at the rate indicated in and up to the limit set out in the Appendix to the Form of Tender upon the amount due to the Contractor on account of sub-clauses (1)(a) and (1)(d) of this Clause and

   (b) any payment which shall have become due under sub-clause (6) of this Clause.

Payment of retention

(6) (a) Upon the issue of a Certificate of Substantial Completion in respect of any Section or part of the Works there shall become due to the Contractor one half of such proportion of the retention money deductible to date under sub-clause (5)(a) of this Clause as the value of the Section or part bears to the value of the whole of the Works completed to date as certified under sub-clause (2)(a) of this Clause and such amount shall be added to the amount next certified as due to the Contractor under sub-clause (2) of this Clause.

The total of the amounts released shall in no event exceed one half of the limit of retention set out in the Appendix to the Form of Tender.

(b) Upon issue of the Certificate of Substantial Completion in respect of the whole of the Works there shall become due to the Contractor one half of the retention money calculated in accordance with sub-clause (5)(a) of this Clause. The amount so due (or the balance thereof over and above such payments already made pursuant to sub-clause (6)(a) of this Clause) shall be paid within 14 days of the issue of the said Certificate.

(c) At the end of the Defects Correction Period or if more than one the last of such periods the final date for payment of the remainder of the retention money to be paid to the Contractor is 14 days later notwithstanding that at that time there may be outstanding claims by the Contractor against the Employer.

Provided that if at that time there remains to be executed by the Contractor any outstanding work referred to under Clause 48 or any work ordered pursuant to Clauses 49 or 50 the Employer may withhold payment until the completion of such work of so much of the said remainder as shall in the opinion of the Engineer represent the cost of the work remaining to be executed.

Interest on overdue payments

(7) In the event of
   (a) failure by the Engineer to certify or the Employer to make payment in accordance with sub-clauses (2) (4) or (6) of this Clause or
   (b) any finding of an arbitrator to such effect

the Employer shall pay to the Contractor interest compounded monthly for each day on which any payment is overdue or which should have been certified and paid at a rate equivalent to 2% per annum above the base lending rate of the bank specified in the Appendix to the Form of Tender. If in an arbitration pursuant to Clause 66 the arbitrator holds that any sum or additional sum should have been certified by a particular date in accordance with the aforementioned sub-clauses but was not so certified this shall be regarded for the purposes of this sub-clause as a failure to certify such sum or additional sum. Such sum or additional sum shall be regarded as overdue for payment 28 days after the date by which the arbitrator holds that the Engineer should have certified the sum or if no such date is identified by the arbitrator shall be regarded as overdue for payment from the date of the Certificate of Substantial Completion for the whole of the Works.

Correction and withholding of certificates

(8) The Engineer shall have power to omit from any certificate the value of any work done goods or materials supplied or services rendered with which he may for the time being be dissatisfied and for that purpose or for any other reason which to him may seem proper may by any certificate delete correct or modify any sum previously certified by him. Provided that

(a) the Engineer shall not in any interim certificate delete or reduce any sum previously certified in respect of work done goods or materials supplied or services rendered by a Nominated Sub-contractor if the Contractor shall have already paid or be bound to pay that sum to the Nominated Sub-contractor and

(b) if the Engineer in the final certificate shall delete or reduce any sum previously certified in respect of work done goods or materials supplied or services rendered by a Nominated Sub-contractor which sum shall have been already paid by the Contractor to the Nominated Sub-contractor the Employer shall reimburse to the Contractor the amount of any sum overpaid by the Contractor to the Sub-contractor in accordance with the certificates issued under sub-clausde (2) of this Clause which the Contractor shall be unable to recover from the Nominated Sub-contractor together with interest thereon at the rate stated in sub-clause (7) of this Clause from 28 days after the date of the final certificate issued under sub-clause (4) of this Clause until the date of such reimbursement.

Copy of certificate for contractor
(9) Every certificate issued by the Engineer pursuant to this Clause shall be sent to the Employer and on the Employer's behalf to the Contractor. By this certificate the Employer shall give notice to the Contractor specifying the amount (if any) of the payment proposed to be made and the basis on which it was calculated.

Payment advice
(10) Where a payment under Clause 60(2) or (4) is to differ from that certified or the employer is to withhold payment after the final date for payment of a sum due under the Contract the Employer shall notify the Contractor in writing not less than one day before the final date for payment specifying the amount proposed to be withheld and the ground for withholding payment or if there is more than one ground each ground and the amount attributable to it.

Clause 66 – Avoidance and settlement of disputes

Avoidance of disputes
(1) In order to overcome where possible the causes of disputes and in those cases where disputes are likely still to arise to facilitate their clear definition and early resolution (whether by agreement or otherwise) the following procedure shall apply for the avoidance and settlement of disputes.

Matters of dissatisfaction
(2) If at any time
    (a) the Contractor is dissatisfied with any act or instruction of the Engineer's Representative or any other person responsible to the Engineer or
    (b) the Employer or the Contractor is dissatisfied with any decision opinion instruction direction certificate or valuation of the Engineer or with

any other matter arising under or in connection with the Contract or the carrying out of the Works

the matter of dissatisfaction shall be referred to the Engineer who shall notify his written decision to the Employer and the Contractor within one month of the reference to him.

Disputes

(3) The Employer and the Contractor agree that no matter shall constitute nor be said to give rise to a dispute unless and until in respect of that matter

(a) the time for the giving of a decision by the Engineer on a matter of dissatisfaction under Clause 66(2) has expired or the decision given is unacceptable or has not been implemented and in consequence the Employer or the Contractor has served on the other and on the Engineer a notice in writing (hereinafter called the Notice of Dispute) or

(b) an adjudicator has given a decision on a dispute under Clause 66(6) and the Employer or the Contractor is not giving effect to the decision, and in consequence the other has served on him and the Engineer a Notice of Dispute

and the dispute shall be that stated in the Notice of Dispute. For the purposes of all matters arising under or in connection with the Contract or the carrying out of the Works the word "dispute" shall be construed accordingly and shall include any difference.

(4) (a) Notwithstanding the existence of a dispute following the service of a Notice under Clause 66(3) and unless the Contract has already been determined or abandoned the Employer and the Contractor shall continue to perform their obligations.

(b) The Employer and the Contractor shall give effect forthwith to every decision of

(i) the Engineer on a matter of a dissatisfaction given under Clause 66(2) and

(ii) the adjudicator on a dispute given under Clause 66(6)

unless and until that decision is revised by agreement of the Employer and the Contractor or pursuant to Clause 66.

Conciliation

(5) (a) The Employer or the Contractor may at any time before service of a Notice to Refer to arbitration under Clause 66(9) by notice in writing seek the agreement of the other for the dispute to be considered under the Institution of Civil Engineers' Conciliation Procedure (1994) or any amendment or modification thereof being in force at the date of such notice.

(b) If the other party agrees to this procedure any recommendation of the conciliator shall be deemed to have been accepted as finally determining the dispute by agreement so that the matter is no longer in dispute unless a Notice to Refer to arbitration under Clause 66(6) has

170

been served in respect of that dispute not later than one month after receipt of the recommendation by the dissenting party.

Adjudication

(6) (a) The Employer and the Contractor each has the right to refer a dispute as to a matter under the Contract for adjudication and either party may give notice in writing (hereinafter called the Notice of Adjudication) to the other at any time of his intention so to do. The adjudication shall be conducted under the Institution of Civil Engineers' Adjudication Procedure (1997) or any amendment or modification thereof being in force at the time of the said Notice.

(b) Unless the adjudicator has already been appointed he is to be appointed by a timetable with the object of securing his appointment and referral of the dispute to him within 7 days of such notice.

(c) The adjudicator shall reach a decision within 28 days of referral or such longer period as is agreed by the parties after the dispute has been referred.

(d) The adjudicator may extend the period of 28 days by up to 14 days with the consent of the party by whom the dispute was referred.

(e) The adjudicator shall act impartially.

(f) The adjudicator may take the initiative in ascertaining the facts and the law.

(7) The decision of the adjudicator shall be binding until the dispute is finally determined by legal proceedings or by arbitration (if the contract provides for arbitration or the parties otherwise agree to arbitration) or by agreement.

(8) The adjudicator is not liable for anything done or omitted in the discharge or purported discharge of his functions as adjudicator unless the act or omission is in bad faith and any employee or agent of the adjudicator is similarly not liable.

Arbitration

(9) (a) All disputes arising under or in connection with the Contract or the carrying out of the Works other than failure to give effect to a decision of an adjudicator shall be finally determined by reference to arbitration. The party seeking arbitration shall serve on the other party a notice in writing (called the Notice to Refer) to refer the dispute to arbitration.

(b) Where an adjudicator has given a decision under Clause 66(6) in respect of the particular dispute the Notice to Refer must be served within three months of the giving of the decision otherwise it shall be final as well as binding.

Appointment of arbitrator
President or Vice-President to act

(10) (a) The arbitrator shall be a person appointed by agreement of the parties.

(b) If the parties fail to appoint an arbitrator within one month of either party serving on the other party a notice in writing (hereinafter called the Notice to Concur) to concur in the appointment of an arbitrator the dispute shall be referred to a person to be appointed on the application of either party by the President for the time being of the Institution of Civil Engineers.

(c) If an arbitrator declines the appointment or after appointment is removed by order of a competent court or is incapable of acting or dies and the parties do not within one month of the vacancy arising fill the vacancy then either party may apply to the President for the time being of the Institution of Civil Engineers to appoint another arbitrator to fill the vacancy.

(d) In any case where the President for the time being of the Institution of Civil Engineers is not able to exercise the functions conferred on him by this Clause the said functions shall be exercised on his behalf by a Vice-President for the time being of the said Institution.

Arbitration – procedure and powers

(11) (a) Any reference to arbitration under this Clause shall be deemed to be a submission to arbitration within the meaning of the Arbitration Act 1996 or any statutory re-enactment or amendment thereof for the time being in force. The reference shall be conducted in accordance with the procedure set out in the Appendix to the Form of Tender or any amendment or modification thereof being in force at the time of the appointment of the arbitrator. Such arbitrator shall have full power to open up review and revise any decision opinion instruction direction certificate or valuation of the Engineer or an adjudicator.

(b) Neither party shall be limited in the arbitration to the evidence or arguments put to the Engineer or to any adjudicator pursuant to Clause 66(2) or 66(6) respectively.

(c) The award of the arbitrator shall be binding on all parties.

(d) Unless the parties otherwise agree in writing any reference to arbitration may proceed notwithstanding that the Works are not then complete or alleged to be complete.

Witnesses

(12) (a) No decision opinion instruction direction certificate or valuation given by the Engineer shall disqualify him from being called as a witness and giving evidence before a conciliator adjudicator or arbitrator on any matter whatsoever relevant to the dispute.

(b) All matters and information placed before a conciliator pursuant to a reference under sub-clause (5) of this Clause shall be deemed to be submitted to him without prejudice and the conciliator shall not be called as witness by the parties or anyone claiming through them in connection with any adjudication arbitration or other legal proceedings arising out of or connected with any matter so referred to him.

APPENDIX 8
PART 24 CIVIL PROCEDURE RULES: SUMMARY JUDGMENT

Contents of this part

Scope of this part

24.1 This Part sets out a procedure by which the court may decide a claim or a particular issue without a trial.

Grounds for summary judgment

24.2 The court may give summary judgment against a claimant or defendant on the whole of a claim or on a particular issue if –

    (a) it considers that –

       (i) that claimant has no real prospect of succeeding on the claim or issue; or

       (ii) that defendant has no real prospect of successfully defending the claim or issue; and

    (b) there is no other reason why the case or issue should be disposed of at a trial.

(Rule 3.4 makes provision for the court to strike out $^{(GL)}$ a statement of case or part of a statement of case if it appears that it discloses no reasonable grounds for bringing or defending a claim)

Types of proceedings in which summary judgment is available

24.3 (1) The court may give summary judgment against a claimant in any type of proceedings.

    (2) The court may give summary judgment against a defendant in any type of proceedings except –

(a) proceedings for possession of residential premises against a tenant, a mortgagor or a person holding over after the end of his tenancy; and

(b) proceedings for an admiralty claim in rem.

Procedure

24.4 (1) A claimant may not apply for summary judgment until the defendant against whom the application is made has filed –

(a) an acknowledgment of service; or

(b) a defence,

unless –

(i) the court gives permission; or

(ii) a practice direction provides otherwise.

(Rule 10.3 sets out the period for filing an acknowledgment of service and rule 15.4 the period for filing a defence)

(2) If a claimant applies for summary judgment before a defendant against whom the application is made has filed a defence, that defendant need not file a defence before the hearing.

(3) Where a summary judgment hearing is fixed, the respondent (or the parties where the hearing is fixed of the court's own initiative) must be given at least 14 days' notice of –

(a) the date fixed for the hearing; and

(b) the issues which it is proposed that the court will decide at the hearing.

(Part 23 contains the general rules about how to make an application)

(Rule 3.3 applies where the court exercises its powers of its own initiative)

Evidence for the purposes of a summary judgment hearing

24.5 (1) If the respondent to an application for summary judgment wishes to rely on written evidence at the hearing, he must –

(a) file the written evidence; and

(b) serve copies on every other party to the application,

at least 7 days before the summary judgment hearing.

(2) If the applicant wishes to rely on written evidence in reply, he must –

(a) file the written evidence; and

(b) serve a copy on the respondent,

at least 3 days before the summary judgment hearing.

(3) Where a summary judgment hearing is fixed by the court of its own initiative –

(a) any party who wishes to rely on written evidence at the hearing must –

(i) file the written evidence; and

(ii) unless the court orders otherwise, serve copies on every other party to the proceedings,

at least 7 days before the date of the hearing;

  (b) any party who wishes to rely on written evidence at the hearing in reply to any other party's written evidence must –
      (i) file the written evidence in reply; and
      (ii) unless the court orders otherwise, serve copies on every other party to the proceedings,
      at least 3 days before the date of the hearing.
  (4) This rule does not require written evidence –
      (a) to be filed if it has already been filed; or
      (b) to be served on a party on whom it has already been served.

Court's powers when it determines a summary judgment application
24.6 When the court determines a summary judgment application it may –
      (a) give directions as to the filing and service of a defence;
      (b) give further directions about the management of the case.
      (Rule 3.1(3) provides that the court may attach conditions when it makes an order)

Practice direction – the summary disposal of claims

This practice direction supplements CPR Part 24

Applications for summary judgment under Part 24
1.1 Attention is drawn to Part 24 itself and to:
     Part 3, in particular rule 3.1(3) and (5),
     Part 22,
     Part 23, in particular rule 23.6,
     Part 32, in particular rule 32.6(2).

1.2 In this paragraph, where the context so admits, the word 'claim' includes:
     (1) a part of a claim, and
     (2) an issue on which the claim in whole or part depends.

1.3 An application for summary judgment under rule 24.2 may be based on:
     (1) a point of law (including a question of construction of a document),
     (2) the evidence which can reasonably be expected to be available at trial or the lack of it, or
     (3) a combination of these.

1.4 Rule 24.4(1) deals with the stage in the proceedings at which an application under Part 24 can be made (but see paragraph 7.1 below).

Procedure for making an application
2   (1) Attention is drawn to rules 24.4(3) and 23.6

(2) The application notice must include a statement that it is an application for summary judgment made under Part 24.

(3) The application notice or the evidence contained or referred to in it or served with it must –

(a) identify concisely any point of law or provision in a document on which the applicant relies, and/or

(b) state that it is made because the applicant believes that on the evidence the respondent has no real prospect of succeeding on the claim or issue or (as the case may be) of successfully defending the claim or issue to which the application relates,

and in either case state that the applicant knows of no other reason why the disposal of the claim or issue should await trial.

(4) Unless the application notice itself contains all the evidence (if any) on which the applicant relies, the application notice should identify the written evidence on which the applicant relies. This does not affect the applicant's right to file further evidence under rule 24.5(2).

(5) The application notice should draw the attention of the respondent to rule 24.5(1).

The hearing

4   (1) The hearing of the application will normally take place before a Master or a district judge.

(2) The Master or district judge may direct that the application be heard by a High Court Judge (if the case is in the High Court) or a circuit judge (if the case is in a county court).

The court's approach

4.1   Where a claimant applies for judgment on his claim, the court will give that judgment if:

(1) the claimant has shown a case which, if unanswered, would entitle him to that judgment, and

(2) the defendant has not shown any reason why the claim should be dealth with at trial.

4.2   Where a defendant applies for judgment in his favour on the claimant's claim, the court will give that judgment if either:

(1) the claimant has failed to show a case which, if unanswered, would entitle him to judgment, or

(2) the defendant has shown that the claim would be bound to be dismissed at trial.

4.3   Where it appears to the court possible that a claim or defence may succeed but improbable that it will do so, the court may make a conditional order, as described below.

Orders the court may make

5.1 The orders the court may make on an application under Part 24 include:

(1) judgment on the claim,

(2) the striking out or dismissal of the claim,

(3) the dismissal of the application,

(4) a conditional order.

5.2 A conditional order is an order which requires a party:

(1) to pay a sum of money into court, or

(2) to take a specified step in relation to his claim or defence, as the case may be,

and provides that that party's claim will be dismissed or his statement of case will be struck out if he does not comply.

(Note – the court will not follow its former practice of granting leave to a defendant to defend a claim, whether conditionally or unconditionally.)

Accounts and inquiries

6 If a remedy sought by a claimant in his claim form includes, or necessarily involves, taking an account or making an inquiry, an application can be made under Part 24 by any party to the proceedings for an order directing any necessary accounts or inquiries to be taken or made.

(This paragraph replaces RSC Order 43, rule 1, but applies to county court proceedings as well as to High Cour proceedings. The Accounts practice direction supplementing Part 40 contains further provisions as to orders for accounts and inquiries.)

Specific performance

7.1 (1) If a remedy sought by a claimant in his claim form includes a claim –

(a) for specific performance of an agreement (whether in writing or not) for the sale, purchase, exchange, mortgage or charge of any property, or for the grant or assignment of a lease or tenancy of any property, with or without an alternative claim for damages, or

(b) for rescission of such an agreement, or

(c) for the forfeiture or return of any deposit made under such an agreement,

the claimant may apply under Part 24 for judgment.

7.1 (2) The claimant may do so at any time after the claim form has been served, whether or not the defendant has acknowledged service of the claim form, whether or not the time for acknowledging service has expired and whether or not any particulars of claim have been served.

7.2 The application notice by which an application under paragraph 7.1 is made must have attached to it the text of the order sought by the claimant.

7.3 The application notice and a copy of every affidavit or witness statement in support and of any exhibit referred to therein must be served on the defendant not less than 4 days before the hearing of the application. (note – the 4 days replaces for these applications the 7 days specified in rule 24.5.)

(This paragraph replaces RSC Order 86, rules 1 and 2 but applies to county court proceedings as well as to High Court proceedings.)

Setting aside order for summary judgment

8.1 If an order for summary judgment is made against a respondent who does not appear at the hearing of the application, the respondent may apply for the order to be set aside or varied (see also rule 23.11).

8.2 On the hearing of an application under paragraph 8.1 the court may make such order as it thinks just.

Costs

9.1 Attention is drawn to Part 44 (fixed costs).

9.2 Attention is drawn to the Costs Practice Direction and in particular to the court's power to make a summary assessment of costs.

9.3 Attention is also drawn to rule 43.5(5) which provides that if an order does not mention costs no party is entitled to costs relating to that order.

Case management

10 Where the court dismisses the application or makes an order that does not completely dispose of the claim, the court will give case management directions as to the future conduct of the case.

PART 25 CIVIL PROCEDURE RULES: INTERIM REMEDIES

Contents of this part

Orders for interim remedies

25.1 (1) The court may grant the following interim remedies –

(k) an order (referred to as an order for interim payment) under rule 25.6 for payment by a defendant on account of any damages, debt or other sum (except costs) which the court may hold the defendant liable to pay;

(3) The fact that a particular kind of interim remedy is not listed in paragraph (1) does not affect any power that the court may have to grant that remedy.

(4) The court may grant an interim remedy whether or not there has been a claim for a final remedy of that kind.

Time when an order for an interim remedy may be made

25.2 (1) An order for an interim remedy may be made at any time, including –

(a) before proceedings are started; and

(b) after judgment has been given.

(Rule 7.2 provides that proceedings are started when the court issues a claim form)

(2) However

(c) unless the court otherwise orders, a defendant may not apply for any of the orders listed in rule 25.1(1) before he has filed either an acknowledgement of service or a defence.

(Part 10 provides for filing an acknowledgement of service and Part 15 for filing a defence)

How to apply for an interim remedy

25.3 (1) The court may grant an interim remedy on an application made without notice if it appears to the court that there are good reasons for not giving notice.

(2) An application for an interim remedy must be supported by evidence, unless the court orders otherwise.

(3) If the applicant makes an application without giving notice, the evidence in support of the application must state the reasons why notice has not been given.

(Part 3 lists general powers of the court)

(Part 23 contains general rules about making an application)

Interim payments – general procedure

25.6 (1) The claimant may not apply for an order for an interim payment before the end of the period for filing an acknowledgment of service applicable to the defendant against whom the application is made.

(Rule 10.3 sets out the period for filing an acknowledgment of service)

(Rule 25.1(1)(k) defines an interim payment)

(2) The claimant may make more than one application for an order for an interim payment.

(3) A copy of an application notice for an order for an interim payment must –

(a) be served at least 14 days before the hearing of the application; and

(b) be supported by evidence.

(4) If the respondent to an application for an order for an interim payment wishes to rely on written evidence at the hearing, he must –

(a) file the written evidence; and

(b) serve copies on every other party to the application,

at least 7 days before the hearing of the application.

(5) If the applicant wishes to rely on written evidence in reply, he must –

(a) file the written evidence; and

(b) serve a copy on the respondent,

at least 3 days before the hearing of the application.

(6) This rule does not require written evidence –

(a) to be filed if it has already been filed; or

(b) to be served on a party on whom it has already been served.

(7) The court may order an interim payment in one sum or in instalments.

(Part 23 contains general rules about applications)

Interim payments – conditions to be satisfied and matters to be taken into account

25.7 (1) The court may make an order for an interim payment only if –

  (a) the defendant against whom the order is sought has admitted liability to pay damages or some other sum of money to the claimant;

  (b) the claimant has obtained judgment against that defendant for damages to be assessed or for a sum of money (other than costs) to be assessed;

  (c) except where paragraph (3) applies, it is satisfied that, if the claim went to trial, the claimant would obtain judgment for a substantial amount of money (other than costs) against the defendant from whom he is seeking an order for an interim payment; or

  (d) the following conditions are satisfied –

    (i) the claimant is seeking an order for possession of land (whether or not any other order is also sought); and

    (ii) the court is satisfied that, if the case went to trial, the defendant would be held liable (even if the claim for possession fails) to pay the claimant a sum of money for the defendant's occupation and use of the land while the claim for possession was pending.

  (2) In addition, in a claim for personal injuries the court may make an order for an interim payment of damages only if –

  (a) the defendant is insured in respect of the claim;

  (b) the defendant's liability will be met by –

    (i) an insurer under section 151 of the Road Traffic Act 1988([33]); or

    (ii) an insurer acting under the Motor Insurers Bureau Agreement, or the Motor Insurers Bureau where it is acting itself; or

  (c) the defendant is a public body.

  (3) In a claim for personal injuries where there are two or more defendants, the court may make an order for the interim payment of damages against any defendant if –

  (a) it is satisfied that, if the claim went to trial, the claimant would obtain judgment for substantial damages against at least one of the defendants (even if the court has not yet determined which of them is liable); and

  (b) paragraph (2) is satisfied in relation to each of the defendants.

  (4) The court must not order an interim payment of more than a reasonable proportion of the likely amount of the final judgment.

  (5) The court must take into account –

  (a) contributory negligence; and

  (b) any relevant set-off or counterclaim.

Powers of court where it has made an order for interim payment

25.8 (1) Where a defendant has been ordered to make an interim payment, or has in fact made an interim payment (whether voluntarily or under an order), the court may make an order to adjust the interim payment.

(2) The court may in particular –

(a) order all or part of the interim payment to be repaid;

(b) vary or discharge the order for the interim payment;

(c) order a defendant to reimburse, either wholly or partly, another defendant who has made an interim payment.

(3) The court may make an order under paragraph (2)(c) only if –

(a) the defendant to be reimbursed made the interim payment in relation to a claim in respect of which he has made a claim against the other defendant for a contribution$^{(GL)}$, indemnity$^{(GL)}$ or other remedy; and

(b) where the claim or part to which the interim payment relates has not been discontinued or disposed of, the circumstances are such that the court could make an order for interim payment under rule 25.7.

(4) The court may make an order under this rule without an application by any party if it makes the order when it disposes of the claim or any part of it.

(5) Where –

(a) a defendant has made an interim payment; and

(b) the amount of the payment is more than his total liability under the final judgment or order,

the court may award him interest on the overpaid amount from the date when he made the interim payment.

Restriction on disclosure of an interim payment

25.9 The fact that a defendant has made an interim payment, whether voluntarily or by court order, shall not be disclosed to the trial judge until all questions of liability and the amount of money to be awarded have been decided unless the defendant agrees.

Practice direction – interim payments

This practice direction supplements CPR Part 25

General

1.1 Rule 25.7 sets out the conditions to be satisfied and matters to be taken into account before the court will make an order for an interim payment.

1.2 The permission of the court must be obtained before making a voluntary interim payment in respect of a claim by a child or patient.

Evidence

2.1 An application for an interim payment of damages must be supported by evidence dealing with the following:

(1) the sum of money sought by way of an interim payment,

(2) the items or matters in respect of which the interim payment is sought,

(3) the sum of money for which final judgment is likely to be given,

(4) the reasons for believing that the conditions set out in rule 25.7 are satisfied,

(5) any other relevant matters,

(6) in claims for personal injuries, details of special damages and past and future loss, and

(7) in a claim under the Fatal Accidents Act 1976, details of the person(s) on whose behalf the claim is made and the nature of the claim.

2.2 Any documents in support of the application should be exhibited, including, in personal injuries claims, the medical report(s).

2.3 If a respondent to an application for an interim payment wishes to rely on written evidence at the hearing he must comply with the provisions of rule 25.6(4).

2.4 If the applicant wishes to rely on written evidence in reply he must comply with the provisions of rule 25.6(5).

Instalments

3 Where an interim payment is to be paid in instalments the order should set out:

(1) the total amount of the payment,

(2) the amount of each instalment,

(3) the number of instalments and the date on which each is to be paid, and

(4) to whom the payment should be made.

Compensation recovery payments

4.1 Where in a claim for personal injuries there is an application for an interim payment of damages:

(1) which is other than by consent,

(2) which falls under the heads of damage set out in column 1 of Schedule 2 of the Social Security (Recovery of Benefits) Act 1997 in respect of recoverable benefits received by the claimant set out in column 2 of that Schedule, and

(3) where the defendant is liable to pay recoverable benefits to the Secretary of State,

the defendant should obtain from the Secretary of State a certificate of recoverable benefits.

4.2 A copy of the certificate should be filed at the hearing of the application for an interim payment.

4.3 The order will set out the amount by which the payment to be made to the claimant has been reduced according to the Act and the Social Security (Recovery of Benefits) Regulations 1997.

4.4 The payment made to the claimant will be the net amount but the interim payment for the purposes of paragraph 5 below will be the gross amount.

Adjustment of final judgment figure

5.1 In this paragraph 'judgment' means:
(1) any order to pay a sum of money,
(2) a final award of damages,
(3) an assessment of damages.

5.2 In a final judgment where an interim payment has previously been made which is less than the total amount awarded by the judge, the order should set out in a preamble:
(1) the total amount awarded by the judge, and
(2) the amounts and dates of the interim payment(s).

5.3 The total amount awarded by the judge should then be reduced by the total amount of any interim payments, and an order made for entry of judgment and payment of the balance.

5.4 In a final judgment where an interim payment has previously been made which is more than the total amount awarded by the judge, the order should set out in a preamble:
(1) the total amount awarded by the judge, and
(2) the amounts and dates of the interim payment(s).

5.5 An order should then be made for repayment, reimbursement, variation or discharge under rule 25.8(2) and for interest on an overpayment under rule 25.8(5).

5.6 A practice direction supplementing Part 40 provides further information concerning adjustment of the final judgment sum.

TABLE OF CASES

TABLE OF STATUTORY MATERIALS

INDEX